Deteriorated concrete

Deteriorated concrete

Inspection and physicochemical analysis

Frank Rendell, Raoul Jauberthie and Mike Grantham

Thomas Telford

Published by Thomas Telford Publishing,
Thomas Telford Ltd,
1 Heron Quay,
London E14 4JD.
URL: http://www.thomastelford.com

Distributors for Thomas Telford books are
USA: ASCE Press, 1801 Alexander Bell Drive, Reston, VA 20191-4400, USA
Japan: Maruzen Co. Ltd, Book Department, 3–10 Nihonbashi 2-chome, Chuo-ku, Tokyo 103
Australia: DA Books and Journals, 648 Whitehorse Road, Mitcham 3132, Victoria

First published 2002

Also available from Thomas Telford Books
Deterioration and failure of building materials. R O Heckroodt, 2002 ISBN: 0 7277 3172 6.
Portland cement. 2nd edn. G C Bye, 1999. ISBN: 0 7277 2766 4.
Cement chemistry. 2nd edn. H F W Taylor, 1997. ISBN: 0 7277 2592 0.

A catalogue record for this book is available from the British Library

ISBN: 0 7277 3119 X

Typeset by Gray Publishing, Tunbridge Wells, Kent
Printed and bound in Great Britain by MPG Books, Bodmin, Cornwall

Contents

Contents

Foreword

Concrete is the most available and widely used of all construction materials. It has, and is, serving humanity very well. So well in fact that it often goes unmaintained and becomes taken for granted because it has an assumed permanency. Phrases such as 'putting matters on a concrete foundation' or 'concrete evidence' implies security and surety.

But, as with all forms of engineering, the achievement of practical solutions brings along with them occasional inadequacy. It is the inadequacy that catches the headlines and can distort the more representative picture. in this regard the three authors of this book (the Franco-Anglo Alliance) have acknowledged that concrete may deteriorate as do all building fabrics. However, these authors show that deterioration can be subject to disciplined scrutiny in a manner representative of forensic science. Compiling such evidence assists in making judgements about cause and effect resulting in diminished performance and, as a result, remedial measures can be based on understanding rather than pragmatic flair.

This book endeavours and, I suggest, succeeds in linking these aspects together by introducing science into what many regard as the province of the practising engineer. Such intrusion should be welcomed and assisted by addressing the problems of technical language, difficult physical and chemical concepts that underwrite many of the diagnostic techniques that are used in this field of concrete pathology, e.g. scanning electron microscopy, X-ray diffraction and fluorescence, as well as microprobe analysis.

The authors explain the actions underlying the techniques that assist in the interpretations that can be made on the measurements taken. It is this combination of evidence, reasoning and pragmatism that are the hallmarks of this book.

As with all investigative work, evidence is gathered from site and a methodology of questioning developed. Such evidence is augmented by laboratory testing and the two sets of data act in a complementary way that allows rational conclusions to be drawn. One might be forgiven in thinking that this would be sufficient for remedial action to be taken and/or blame to be apportioned. Not so, the fundamental understanding of the techniques are explained in such a way that allows a better appreciation to be made of the capabilities and, equally important, the limitations of such analytical methods. The combination of science and practice have been brought together very effectively. It is tempting to overcomplicate such specialisms and this the authors have carefully avoided. So, does it all work in practice? The five case histories show well the effectiveness of this qualitative and quantitative forensic approach.

The authors maintain the explanatorily approach adopted in the book to the very end with three Appendices that explain difficult phenomena associated with forms of radiation and solid body responses, crystal structure and data interpretation; the latter being an excellent summary of benchmarking information that permits practical conclusions to be drawn.

I hope this book brings together investigative analysis and pragmatic remedial specifications. It certainly deserves such an outcome from which the issue of deteriorating concrete, its causes and treatments will benefit.

Professor P.C. Hewlett
Chief Executive
British Board of Agrément

———

The authors' expertise and the quality of the themes developed in each section means that this book will provide extensive information and best practice to engineers and students or young researchers on concrete durability and techniques devoted to deterioration diagnosis. It deserves to be a best-seller in its domain.

Jean-François Coste
Past President of the French Civil Engineering Association
AFGC (Association Française de Génie Civil)

Preface

As the infrastructure asset base ages, engineers have to address the problems associated with deterioration. Concrete is one of the most commonly used materials and yet its properties and behaviour in service are often shrouded under the cloak of chemistry and petrography. This book has been written for engineers to help to guide them through the morass of techniques they may encounter when involved in the evaluation of concrete structures. One of the aims of this book is to give an insight into the physicochemical analytical techniques that are now available to the engineer and concrete scientist. Equally, this text will provide a good starting point for students and researchers working in the field of advanced concrete technology.

The book contains a body of information concerning the fundamentals of concrete technology directly applicable to the diagnosis of problems with concrete. The reader is then introduced to *in situ* and preliminary laboratory testing of concrete. The final sections discuss the application of X-ray diffraction and scanning electron microscopy to concrete technology. The content of the book is then underpinned by a number of case studies to illustrate the practical problems of concrete inspection and analysis.

The three authors come from diverse backgrounds and therefore have been able to contribute a wide range of perspectives.

Dr Frank Rendell is an independent researcher and member of the Laboratory GRGCR (Groupe de Researche Génie Civil, Rennes), INSA (Institut National de Sciences Appliquées), Rennes, France, and a visiting lecturer at INSA, Rennes. He has research and industrial experience in assessing the behaviour of concrete in extreme environments, including underwater concrete inspection and asset evaluation in wastewater systems.

Dr Raoul Jauberthie is a Maître de Conférences, Départment Génie Civil et Urbanisme, member of the Laboratory LMSM (Laboratoire de mécanique des structures et matériaux) at INSA, Rennes. He has a background in material science and physics. He specializes in the study of concrete behaviour under aggressive conditions and the development of sustainable materials. This has included work for developing countries relating to the exploitation of waste products as admixtures. Dr Jauberthie has also made a notable contribution to the application of scanning electron microscopy and X-ray diffraction analysis for the evaluation of concrete.

Mike Grantham is a Director of MG Associates Construction Consultancy Ltd. He is a European Chemist (EurChem) and a Fellow of the Royal Society of Chemistry. He has worked for several of the major concrete testing houses and has many years experience in the investigation of concrete problems. He is a well-known lecturer in the field of diagnosis, inspection and repair of concrete structures.

Préface

Au fur et à mesure que les ouvrages vieillissent, les ingénieurs sont de plus en plus confrontés à leurs détériorations. Le béton est un matériau très couramment utilisé mais ses propriétés peuvent être gravement affectées dans des environnements chimiquement pollués. Ce livre devrait servir de guide pour des ingénieurs en leur présentant les techniques à leur disposition pour évaluer l'état des structures en béton. Ce livre est une première approche des techniques d'analyse physico-chimiques pour des ingénieurs et scientifiques. Le texte doit donner un bon point de départ à des étudiants ou à des chercheurs dans le domaine des bétons.

Cet ouvrage est une bonne source d'informations sur la technologie des bétons. Il permet de faire un bon diagnostic des problèmes qui sont liés à ce matériau. Il aborde également les tests de laboratoire avec des applications de la diffraction X et de la microscopie électronique à balayage. Le contenu de ce livre est illustré par des examples précis de dégradations observés sur des bâtiments ou sur des ouvrages d'art. Ces diagnostics une fois établis, sont suivis de propositions pour une remise en état.

En conclusion de cette préface, Jean-François Coste, Président de l'Association Française de Génie Civil, apporte un avis éclairé sur ce livre dans les termes suivants:

———

Merci aux auteurs pour la qualité des thèmes développés dans chaque chapitre de ce livre. Il fournit des informations très pratiques à des ingénieurs, étudiants ou jeunes chercheurs dans le domaine de la durabilité des bétons et des techniques d'études utilisables pour diagnostiquer les dégradations. Il devrait être le *best-seller* dans ce domaine.

Jean-François Coste
Président de l'Association Française de Génie Civil

Acknowledgements

The authors wish to express their thanks to the many people who have helped in the preparation of this book. Particular thanks is extended to the following.

Mike Eden of Geomaterials Research Services Ltd, Essex, for his advice and guidance on concrete analysis.

Sandrine Garnier, Groupe de Researche en Chimie et Métallurgie (GRCM) INSA, Rennes, for her assistance with X-ray diffraction analysis.

M Lelanic, CMEBA Université de Rennes I for his assistance with scanning electron microscopy.

Highways Agency and Halcrow for their permission to reproduce the photograph of thaumasite damaged concrete, Figure 3.10.

Finally, we wish to thank GKM without whose support and encouragement this book would not have been possible.

Dedication

To GKM without whose support and encouragement this book would not have been possible.

Chapter 1

Introduction

Anybody involved in the maintenance of reinforced concrete structures will be aware of the substantial increase in the amount of repair work now being carried out. This may merely reflect the fact that a large number of reinforced concrete structures are now coming of age, although certain changes in materials over the past 20 years have been identified as being partially responsible for some of the problems.

For example, in the early 1970s it was believed that there were no deposits of alkali-reactive aggregates in the UK and yet within a few years alkali aggregate reactions became a subject of major interest as more and more cases came to light. This upsurge in alkali silica reaction (ASR) problems may be partially traceable to an increase in cement alkali content in the preceding years. Furthermore, increases in the early-age strength of cement may have meant that specified minimum strengths were being achieved with less cement than had previously been used, with a corresponding effect on subsequent durability. In the light of such phenomena it is clear that concrete must be regarded as a complex mixture of materials, each component of which may in itself vary or be affected by environmental factors.

From personal experience, 90% of the problems that will be experienced in concrete repair will involve steel reinforcement corrosion as a primary problem. For the most part this will have been caused entirely by simple carbonation/low cover and/or the presence of chloride salts, either from calcium chloride used as an accelerator or from de-icing salt.

However, in some cases, other more subtle defects may be present, such as shrinkable aggregates, ASR, frost attack, sulfate attack, structural cracks or a whole variety of other possibilities. It must be borne in mind that many of these phenomena may reveal themselves first in areas of low cover and carbonated concrete, perhaps because micro-cracking from one or other of these causes has permitted carbonation to advance more rapidly than might otherwise have been the case. In such cases, it is all too easy to look at the effects of the problem, i.e. steel corrosion, and attribute this as the cause. Attempting to repair concrete affected by such problems may simply mean that the problem recurs in a relatively short space of time.

Durability

A definition of durability

The durability of a material can be thought of as its ability to withstand environmental deterioration. As engineers, one of our primary interests in durability is the deterioration of physical properties of the material; the rate at which these properties deteriorate is directly linked to the service life of the structure. Durability can be viewed in two ways, from a global and from a material performance view. The global view of durability, normally taken in project planning, consists of the deterioration of the functionality of the project through its life. Figure 1.1 illustrates the effect of life-cycle costing of a project; in the ideal case, the life cost of the project will consist of the capital cost and routine maintenance. When problems arise, interventions are called for to maintain the functionality of the project.

A more specific view of durability is from the material behaviour perspective. The durability of specific materials is normally defined in terms of the time taken for various material properties to deteriorate to a specific limit, e.g. the time taken for the strength of a concrete to reach a value assumed at the design stage.

Designing for durability

Designing for durability is the process in which the following assessments are made.

- Define the service life of the structure.
- Plan the finances of the project.

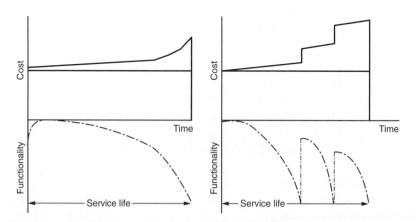

Figure 1.1 *Illustration of the life costing of a structure. On the left is the ideal case in which the structure gradually suffers deterioration with no major need for repair. The case shown on the right illustrates the impact of premature deterioration that seriously reduces the functionality to a point where major works have to be carried out to maintain service.*

- Identify the nature and severity of the environment.
- Investigate the site.
- Identify the mechanisms of attack that will cause material damage.
- Assess the rate at which material properties will deteriorate.
- Assign exposure classifications, e.g. EN 206.
- Select materials and/or provide protection to ensure that the design life can be achieved.
- Ensure the control of material quality and standards of construction.
- Propose an inspection programme.

Assessment of performance

Once the structure has been designed to resist the imposed loads and environmental aggression, it may be found that the performance deviates from that envisaged at the design stage. The aim of this assessment will be to determine the existing degree of functionality and the rate at which the project is deteriorating, and to estimate the revised service life of the project. Routine inspection programmes of structures identify and monitor the evolution of performance; however, all too often assessment is called for when a problem comes to light. Where defects are found it is the role of the engineer to re-evaluate the performance of the structure and if necessary to propose remedial works.

In general terms, the assessment of functionality of a structure will centre around the following framework.

- Determine the aggressivity of the environment.
- Determine the current condition of the materials.
- Define the current level of safety and functionality.
- Evaluate future usage of the structure, revised service life, reduced technical performance levels, etc.

Material performance is judged by *in situ* site assessment and laboratory testing aimed at finding the current performance levels. These studies will be used to confirm that the functionality of the material is acceptable or, in the case of deterioration, they will be used to diagnose the cause of the deterioration to enable a revision of material performance levels. This type of assessment is inevitably complex and the evaluation team often looks to a wide range of techniques to propose and verify their findings.

The aim of this book

The aim of this book is to provide an overview of the assessment of deteriorated concrete, to discuss certain techniques in detail and to demonstrate an application of certain techniques by the use of case studies.

The first part of the book covers some aspects of concrete technology and some of the more common mechanisms of deterioration.

Methods of material evaluation are addressed in two parts: *in situ* site testing and laboratory-based studies. One of the principal aims is to explain the use of certain physicochemical analytical techniques. The areas discussed include the use of X-ray diffraction analysis, scanning electron microscopy and micro-probe analysis. To most civil and structural engineers the complexity of these rather exotic techniques is offputting, to say the least. It is hoped that the presentation of the book will enable an engineer, with a basic knowledge of physics and chemistry, to gain an appreciation of the capabilities and limitations of these analytical methods. The final sections of the book discuss case studies that illustrate the application of techniques developed previously.

Certain topics, such as crystallography and the physics behind X-ray diffraction and electron microscopy, have been set out in appendices. An in-depth knowledge of the information set out in these appendices is not essential, but the reader is advised to have at least a passing knowledge of the terminology.

The book has been written by a French/English team, therefore references are cited from the bodies of literature published in both languages.

Chapter 2

Concrete

Composition and structure of concrete

To understand the analysis of deteriorated concrete it is necessary to have a good appreciation of the nature of sound concrete. The objectives of the chapter are:

- to describe the basic components of concrete: cement, pozzolans and aggregates
- to describe the composition and morphology of a hydrated concrete
- to describe the structure of hydrated concrete.

It is expected that readers have a basic knowledge of concrete technology. If this is not the case they are referred to references 1–5.

Notation

Throughout this book the following shorthand notation is used to describe compounds:

A	Al_2O_3
C	CaO
H	H_2O
F	Fe_2O_3
f	FeO
M	MgO
S	SiO_2
s	SO_4

Constituents of concrete

Cements and pozzolans

For background reading, see references 6–8.

This section presents an overview of materials that undergo chemical and physical transformation to produce the cementitious products used in concrete. Broadly speaking, the materials used in the cement phase of a concrete can be divided into two groups: cementitious materials (hydraulic cements) and pozzolanic materials.

A cementitious material is one that reacts with water to form a solid matrix, whereas pozzolans will not dissolve in or react with water but undergo a slow chemical transformation in the presence of water and lime.

Cements and certain pozzolans are products of the fusion of oxides at high temperatures. The three principal oxides involved in this process are silica, SiO_2, alumina, Al_2O_3, and lime, CaO. By varying the relative ratios of these oxides in the fusion process it is possible to produce products with differing properties. In the case of cements, the cooling of the fused oxides is relatively slow and crystalline compounds are formed as a clinker, which is finely ground into cement. A pozzolan is strictly speaking a mineral admixture; generally, these materials are either naturally occurring or by-products of a process that involves combustion or smelting. In the case of these materials the cooling after the fusion of the oxides is rapid and predominantly amorphous products are formed.

Chemical composition

The chemical composition of cements and pozzolans can be described in terms of their oxide contents (Figure 2.1). The pozzolans tend to be rich in silica and the hydraulic cement materials much richer in calcium oxide. Table 2.1 gives an indication of the range of oxides found and their relative proportions.

Figure 2.1 illustrates that a variation in the ratio of oxides determines the type and therefore properties of a cementitious material. It can be seen from Table 2.1 and Figure 2.1 that the principal difference between the

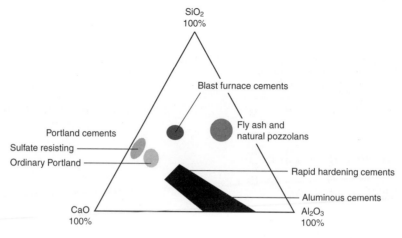

Figure 2.1 *Relationship between the relative proportion of oxide and cement type.*

cementitious materials and the pozzolans is the low lime content in the pozzolans.

Portland cements
Portland cement is produced by the fusion of two raw materials, one rich in lime and the other rich in silica. The raw materials are ground and heated to high temperature (about 1450°C) where they decompose and form, on cooling, a clinker that is composed of various crystalline compounds.

During the formation of the clinker, five principal components are present: CaO, MgO, Al_2O_3, Fe_2O_3 and SiO_2. These oxides mutually react in the following fashion.

- The Fe_2O_3 reacts with the Al_2O_3 and CaO to form the ferrite or Brownmillerite phase C_4AF.
- The Al_2O_3 not used in this reaction reacts with CaO to form tricalcium aluminate, C_3A.
- The remaining lime, CaO, then reacts with the SiO_2 to form belite, C_2S.
- In the presence of excess lime, the belite and lime form alite, C_3S.

The magnesium oxide is not involved in these reactions.

The composition of cement is often described in terms of an oxide analysis (Table 2.1), but a more useful description is given in terms of the crystalline compounds. By varying the ratio of the principal oxides in the cement manufacture the crystalline composition of the clinker can be controlled. The proportion of these compounds dictates the properties of the cement and is used in the classification of cement type. Table 2.2 shows a broad classification of cement type in terms of the principal compounds.

Table 2.1 *Typical oxide composition (% by weight)*

Oxide	Cementitious			Pozzolans			
	OPC	SRC	AC	PFA	Pozzolana	Silica fume	GGBFS
SiO_2	21	25	10	48	65	93	38
CaO	63	65	40	5	4	<1	38
Al_2O_3	5	2	40	25	13	<1	10
Fe_2O_3	3	5	10	10		1	<1
MgO	2	2	1	2		<1	5
Others				10	18		

OPC: ordinary Portland cement; SRC: sulfate-resisting Portland cement; AC: aluminous cement; PFA: pulverized fuel ash; GGBFS: ground-granulated blast furnace slag.

Cement quality

Cement producers monitor the quality of their products, and the laboratory chemical reports and cement test reports are very useful documents when assessing the properties and quality of a cement. Tables 2.3 and 2.4 set out an example of these reports with an explanation of the significance

Table 2.2 *Typical compound composition of Portland cements (% by weight)*

Cement type	C_3S	C_2S	C_3A	C_4AF	Gypsum	Others
Portland	50	25	12	6	3	4
Rapid hardening Portland	55	16	12	8	4	5
Low heat Portland	30	45	5	13	3	4
Sulfate resisting	44	35	4	12	3	2

Table 2.3 *Example of a laboratory chemical report for a CEM I 42.5 N cement. The limits indicated in the table relate to a strength grade 42.5 N cement*

Compound	Typical value (%)	Limit	Signification
SiO_2	20.83		
Al_2O_3	4.88		Principal oxides
Fe_2O_3	2.98		
CaO	64.54		
MgO	2.12	<5	MgO (periclase) can cause a reduction in the rate of hydration and can cause swelling in the cement paste (unsoundness). Normal range of MgO: 1–3.5%
SO_3	2.93	<3.5	SO_3 content is a reflection of the gypsum content added to control the set. Limit typically <4%
K_2O	0.62		See total alkalinity
Na_2O	0.28		
Cl	0.04	<0.1	High chlorides can contribute to corrosion problems in reinforcement
Loss on ignition	0.63	<5	Indication of the freshness of a cement. Limit <5%
Non-detectable	0.15		
Insoluble residue	0.38	<5	Normally quartz and some alumina and iron oxides. Normal limits <2–4% depending on the grade of cement
Free CaO	1.7		Free lime, a result of incomplete fusion or overloading of lime in the production process. Excessive free lime can give rise to delayed swelling in the cement paste (unsoundness). Normal range of free lime is 0.5–1.5%

Table 2.3 *(Continued)*

Compound	Typical value (%)	Limit	Signification
Total alkali	0.69		High alkali content may give rise to problems with reactive aggregates (alkali aggregate reaction). It is expressed as equivalent sodium oxide: total equivalent alkali (Na_2Oe) = K_2O + 0.658 Na_2O
Lime saturation factor (LSF)	0.95		The ideal condition is LSF = 1; all the lime is consumed in the production of clinker. A high LSF implies a high free lime content. Normal range 0.6–1.02%
Cement compounds by calculation			Calculation carried out by the Bogue equations
C_3S	53.5		High C_3S content leads to rapid strength gain
C_2S	18.9		High C_2S content is used for low heat cements
C_3A	7.8		Sulfate-resisting cements limit C_3A content: typically <5%
C_4AF	9.0		

Table 2.4 *Cement test report for a CEM I 42.5 N. The limits indicated in the table relate to a strength grade 42.5 N cement*

Property		Typical value	Limit	Signification
Fineness	Specific surface m²/kg	375		The finer the cement the faster the reaction rates. Cements and pozzolans typically have a fineness of the order of 380 m²/kg
Setting time	min	110	>60 min	
Soundness	mm	1	<10 mm	The soundness, a measure of the early age expansion of a mortar, reflects the volumetric stability of a concrete
Compressive strength	2 days N/mm²	33	>20 N/mm²	The two day strengths are used to specify the early-age strength classification of the cement
Compressive strength	28 days N/mm²	50	>42.5 <62.5 N/mm²	

of various results. The chemical composition consists of an oxide analysis followed by the calculated values of the cement compounds. The cement properties monitored by a manufacturer indicate conformity with cement standards; these standards would typically include limits of strength, initial setting time, soundness and a number of chemical properties. Types of cement are standardized to EN 197-1; the strength classes of the cements relate to 28 day standard strength grades: $32.5\,N/mm^2$, $42.5\,N/mm^2$ and $52.5\,N/mm^2$. An additional strength classification, N (normal) and R (rapid), relates to early-age strength.

Aluminous cements

Aluminous cements are produced by the fusion of aluminous and calcareous materials. They have been in use since early in the twentieth century and are noted for their sulfate-resisting properties and their rapid strength gain (up to 80% of the final strength in 24 h). Ciment fondu is one of the most widely used aluminous cements and is often used for rapid repair work. High alumina cements (HAC), containing up to 80% monocalcium aluminate, were associated with a series of failures in the 1970s and this was attributed to a chemical conversion taking place in the hydrated phases. In several of the failures, however, conversion was only partly responsible; other factors, such as detailing of bearings, were also partly responsible. The principal constituents of the unhydrated cement are shown in Table 2.5. The mineralogical composition of an aluminous cement will comprise a large proportion of monocalcium aluminate (CA), plus a group of other compounds such as calcium aluminoferrites (C_4AF), calcium disilicate (βC_2S) and oxides of iron (FeO).

Pozzolans

Pozzolans occur naturally or can be produced from certain waste materials; unlike clinker they are principally amorphous materials and are generally rich in silica (Figure 2.1). Pozzolans are generally used as a partial replacement for cement and these products are often termed blended cements. The general advantages of using pozzolans are cost and improved durability, by virtue of reducing the Portlandite content in the cement paste, and by decreasing permeability. If a pozzolan is exposed to

Table 2.5 *Typical values of an oxide analysis of ciment fondu (% by weight)*

Main constituents	%	Minor constituents	%
Al_2O_3	>37.0	TiO_2	<4.0
CaO	<39.8	MgO	<1.5
SiO_2	<6.0	SO_3	<0.4
Fe_2O_3	<18.5	$K_2O + Na_2O$ soluble	<0.4

water no cementitious process takes place; however, the hydration of a pozzolan results from a reaction between lime and amorphous silica, which is sometimes known as latent hydraulicity. The most reactive pozzolans are associated with fine particles of amorphous silica. Table 2.6 sets out some examples of pozzolanic materials.

Aggregates

For background reading, see references 9–11.

The aggregate accounts for about three-quarters of the total volume and has to fulfil two basic requirements: first, it should not adversely affect the properties of the fresh or hardened concrete, and secondly, it should be chemically and physically stable during the service life of the concrete. The aggregate source will frequently be a compromise between cost and properties; transport being a dominant element in aggregate cost. Within a concrete mix it is often necessary to use different sources for the fine and course aggregates, therefore in assessment of properties and performance of a concrete both sources must be considered. The sources of aggregates can be grouped as follows:

- sea-dredged aggregates   quartz sands and chert or flint aggregate
- quarried and crushed rock   a vast range of different geological types of material
- excavated sands and gravels   quartz sands and chert or flint aggregate
- waste materials and by-products   blast furnace slag
- manmade aggregates   lightweight aggregates such as lytag or perlite
- recycled aggregates   crushed concrete

In the selection of an aggregate for a concrete mix the physical and mineralogical properties must be considered. The physical nature of an aggregate will be described by its mechanical strength, its porosity, the

Table 2.6 *Examples of pozzolanic materials used in concrete*

Nature	Material	Examples
Natural pozzolans	Volcanic ash	Trass, pozzolana
	Lava	Volcanic tuff, pumice
Artificial pozzolans	Waste products	Fly ash, pulverized fuel ash (PFA) Ground-granulated blast furnace slag (GGBS) Silica fume
	Organic wastes	Rice husk ash
	Calcinated clays	Metakaolin

maximum and minimum particle size, the particle shape and texture, and the grading of the mixture. The ideal aggregate will be well graded and have the lowest surface area per unit volume, and the aggregate mixture should have the lowest possible void content. The particle density is also a fundamental consideration in the aggregate selection for lightweight and heavyweight concretes. Problems associated with an inappropriate aggregate will be discussed in Chapter 3.

Composition of hydrated concrete

Hydration

A knowledge of the processes involved in the hydration of a cement is an essential foundation to an understanding of the composition and behaviour of a hardened concrete. As discussed above, a cement can be described in terms of its component oxides, the three principal oxides being calcium oxide, CaO, alumina, Al_2O_3, and silica, SiO_2. The relative ratio of these compounds within a cement will control the reaction products formed upon hydration. Figure 2.2 presents a simplified representation of the relationship between these oxides and their hydrated products in the presence of gypsum. The figure indicates cementitious material: clinker (Portland cements) and aluminous cements. For reference the pozzolans are also shown: blast furnace slag, fly ash and natural pozzolans. The hydration of these cementitious materials leads to the formation of the hydration products indicated at the apexes of the geometric figure that surrounds the point representing the composition.

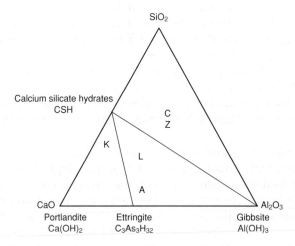

Figure 2.2 *Reaction products expressed in terms of oxide concentration. K: clinker; A: aluminous cements; L: blast furnace slag; C: fly ash; Z: natural pozzolans.*

Hydration of Portland cements[12]

From Figure 2.2 it can be seen that three reaction products will be formed. To gain a deeper insight into the formation of the products, consider the hydration of each phase in the cement.

Hydration of alite (C_3S, $3CaO \cdot SiO_2$)
Reaction products:

CH	$Ca(OH)_2$	Portlandite
CSH(II)	$2CaO \cdot SiO_2 \cdot n\, H_2O$	(n = 2–4)
CSH(I)	$CaO \cdot SiO_2 \cdot m\, H_2O$	(m = 0.5–1.3)

CSH(I) and CSH(II) are sometimes referred to by a generic name, Tobermorite or tobermorite gel. This term covers a range of subspecies of the phase; these subspecies can be characterized by the CaO/SiO_2 ratio of the phase.

The hydration of alite gives rise to the formation of Portlandite and CSH; this hydration is responsible for the relatively rapid development of the strength of the cement paste. The two calcium silicate hydrates formed are CSH(II) and CSH(I). The nearly amorphous CSH(II) is the main cementing phase for concretes cured below 60°C. The CaO/SiO_2 ratio for CSH(II) is in the range of 1.5–2.0 with a normal value of about 1.8; however, in cement pozzolan blended cements this ratio may be lower. The CSH phases produced at elevated temperatures tend to be more crystalline in nature; this trend towards increased crystallinity can sometimes be seen in concrete exposed to fire. The morphology of the calcium silica hydrates is described in Chapter 8.

Hydration of belite (βC_2S, $2CaO \cdot SiO_2$)
Several species of the C_2S phase exist but only βC_2S hydrates in the presence of water. The same reaction products are formed for belite as for alite, but a smaller quantity of Portlandite is formed. This hydration reaction is, as in the case of alite, responsible for developing the strength in the cement paste; however, the hydration process is much slower in this case, typically 5% complete after 1 day, whereas the hydration of alite is typically 40% complete in the same period. The low evolution of Portlandite, as will be discussed later, is of great benefit in cements exposed to chemical aggression.

Hydration of tricalcium aluminate (C_3A, $3CaO \cdot Al_2O_3$)
Tricalcium aluminate undergoes a rapid hydration, typically 80% complete after 1 day, which has a major impact on the setting and early-age strength development in the cement paste. However, the overall contribution of C_3A to the strength of the cement is minor. Gypsum is normally blended with clinker to prevent the 'flash' setting of the C_3A. The gypsum reacts with the C_3A to form ettringite:

$$C_3A\overline{S}H_{32} \qquad\qquad 3CaO\, Al_2O_3\, 3CaSO_4\, 32H_2O$$

This primary ettringite coats the unhydrated C_3A and slows down the set. With time there is a conversion of the primary ettringite to monosulfate:

$$C_3AsH_{12} \qquad 3CaO\ Al_2O_3\ CaSO_4\ 12H_2O$$

The hydration of the tricalcium aluminate can also take place in the absence of gypsum. In this case the hydration reaction yields tricalcium aluminate hexahydrate:

$$C_3AH_6 \qquad 3CaO\ Al_2O_3\ 6H_2O$$

If sulfate ions enter the cement matrix of a mature concrete it is possible to convert the monosulfate back to the expansive ettringite. The formation of this phase has a potentially disruptive effect on the concrete matrix. To limit this effect C_3A contents of cements are limited when the concrete is to be subjected to sulfate conditions. In addition to having a low C_3A content, sulfate-resisting cements are associated with a high C_3S content; the implication of this is that there is a rapid strength gain, which may limit early-age cracking and thus contribute to its resistance to chemical attack.

Hydration of Brownmillerite (C_4AF, $4CaO \cdot Al_2O_3 \cdot Fe_2O_3$)
Brownmillerite and other calcium aluminate ferrites hydrate in a similar fashion to calcium aluminates: a rapid hydration with a minor contribution to ultimate strength. These reaction products, unlike those of C_3A, do not undergo swelling in the presence of sulfates.
   The reaction products of Brownmillerite are:

$C_4A \cdot 13H_2O$	Tetracalcium aluminate hydrate	Crystalline
$C_4F \cdot 13H_2O$		
$(AF)H_3$	Ferric aluminium hydroxide	Amorphous

Hydration of aluminous cement (CA, $CaO \cdot Al_2O_3$)
The hydration of the monocalcium aluminate CA is a rapid reaction responsible for the rapid strength gain. Other compounds such as βC_2S react more slowly and contribute to the ultimate strength. The principal hydration product of aluminous cement is hydrated calcium mono-aluminate, CAH_{10}, and small quantities of dicalcium aluminate 8-hydrate C_2AH_8 are also formed. The main hydrate is meta-stable and under certain conditions it will change to the more stable tricalcium aluminate hex-hydrate (hydrogarnet), C_3AH_6, and gibbsite, AH_3.
   The problem of 'conversion' may occur at normal temperatures but is accelerated by high temperatures and moist conditions, and results in a loss in strength and an increase in porosity.

Hydration of pozzolans
As discussed above, pozzolans are generally amorphous compounds high in silica. To enable the formation of cementitious products from a pozzolan,

Table 2.7 *Hydration products resulting from the reaction between lime and pozzolans*

Blast furnace slag	+Lime	CSH	Tobermorite gel
		C_4AH_{13}	Tetracalcium aluminate hydrate
Blast furnace slag	+Lime	CSH	Tobermorite gel
	+gypsum	C_3AsH_{32}	Ettringite
Natural pozzolan	+Lime	CSH	Tobermorite gel
and PFA	+Lime	C_4AH_{13}	Tetracalcium aluminate hydrate

a source of lime must exist, which can be supplied by the addition of lime to the mix, or by blending the pozzolan with a Portland cement. During the hydration of a blended cement the initial hydration follows the course of normal Portland cement; the liberated Portlandite then reacts with the pozzolan to produce hydration products, notably CSH. Thus, the blending of a pozzolan will have the effect of reducing the Portlandite (an effect beneficial to durability) and increasing the CSH content in the hydrated cement matrix. This hydration of the pozzolans is a slow reaction and accounts for the slow strength growth associated with pozzolanic cements. Examination of a pozzolanic cement will also inevitably reveal quantities of unreacted pozzolan.

The reaction products due to the hydration of pozzolans are dependent on the relative ratios of the oxide composition. Figure 2.2 shows the natural composition of three pozzolans (marked L, Z and C). In the case of a blend of pozzolan and Portland cement the initial reaction occurs in the region of K. As the pozzolans react with the liberated Portlandite the zone describing the reaction products is driven to the right, thereby reducing the $Ca(OH)_2$ content and changing the nature of the hydration products. A more detailed description of the possible reaction products is shown in Table 2.7.

Dynamics of hydration

From the moment when the unhydrated cement comes into contact with water the process of hydration begins. During the initial phase hydration products begin to form around the cement particles; while these particles remain separate the concrete is in a fluid state. As the layer of hydration products builds up, bridging begins to occur and the particles begin to form a loose structure; this continues until the structure develops a fixed form. This period is generally characterized by the initial and final set of the concrete. Once the cement paste has hardened, the growth of CSH hydration products continues and calcium hydroxide crystals gradually fill the voids between hydrated particles. Figure 2.3 shows schematically the evolution of hydration products with respect to the porosity of the cement paste.

As can be seen from Figure 2.3, there is a systematic reduction in the porosity of the cement paste as the CSH and Portlandite fill the structure.

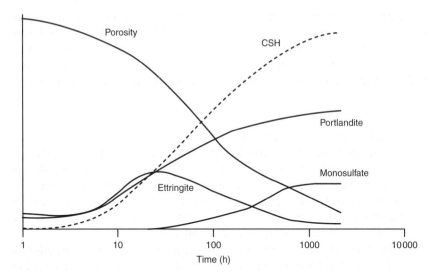

Figure 2.3 *Evolution of the reaction products during hydration.*

Another interesting trend shown in this illustration is the dynamic relationship between the presence of ettringite and monosulfate.[13] As discussed above, there is a conversion of the primary ettringite to monosulfate. In practice, ettringite is potentially harmful owing to its expansive nature. The formation of ettringite can be suppressed by curing the concrete at a high temperature: ettringite is unstable at high temperatures and the monosulfate phase is formed in preference. However, ettringite may then be formed later (delayed ettringite formation) and, in the presence of moisture, can cause damaging expansion to the hardened concrete.

At the end of the period of cure the CSH component of the matrix consists of a porous structure of an amorphous material and occupies about 55% of the volume of the matrix. The crystalline calcium hydroxide, Portlandite, occupies about 20% of the volume and the calcium sulfoaluminates about 10% of the volume.

Chemical admixtures

Throughout history various materials have been added to concretes and mortars in an attempt to improve their performance; for example, the Chinese added eggs and rice to mortars and the Romans used bulls' blood. An admixture is a material that is added to the mix to improve or modify its behaviour, i.e. the handling and setting properties of a fresh concrete (accelerators, retarders, plasticizers, etc.), or to modify the long-term behaviour of the material (air entraining agents, water repellents, etc.). In many countries calcium chloride was used to accelerate the set of concrete when placing concrete in cold weather. The ensuing catalogue of cases of corrosion to reinforcement served to make engineers cautious about the use of admixtures. The development of admixtures and research into secondary effects has now led to their wide acceptance in construction practice. Admixtures can be classified into two groups: mineral admixtures and chemical admixtures. For background reading, see references 14 and 15.

Modification to the set and hydration
- **Accelerators** are admixtures that provoke a fast set in the concrete and a rapid strength growth. They are often used when placing concrete in cold conditions to prevent freeze–thaw damage to the green concrete. A cement paste with an accelerator will typically achieve a final set within 1 h. A concrete with an accelerator can increase the 3 day strength typically by 25%.
- **Retarders** are used to delay the set of a concrete, and consequently delay the hardening. They are used when placing concrete in hot climates and/or when there is an excessive time lag between mixing and placing. A cement paste with a retarder will typically achieve a final set in 3 h. A concrete with a retarder will typically reduce the 7 day strength by 20%.

Modification to workability and properties
- **Plasticizers:** the workability of concrete is a vital factor in concrete construction; to produce concrete that will ensure ease of placing and good compaction requires a fluid mix. The simplest way to achieve this is to increase the water content of the mix; this will have an adverse effect on quality, as will be discussed in the following section. A good workability with a low water content can be achieved by the use of plasticizers.
- **Freeze–thaw resistance:** a concrete submitted to cycles of freeze–thaw will suffer progressive cracking due to the expansion of water on freezing. Since the 1930s air has been entrained into the concrete mix to reduce freeze–thaw damage. The air content of a typical concrete is of the order of 2%; the use of an air entraining agent will elevate this to 5–7%. The voids produced in the concrete range in diameter from 50 to 150 μm. An other advantage of air entrainment is the improve-ment in the rheology of the fresh concrete, i.e. it increases slump, and reduces segregation and bleeding.
- **Waterproofing:** the objective of these admixtures is to reduce the per-meability of the concrete; this covers the cases of capillary absorption and saturated flow in concrete. These admixtures function by block-ing the pore and capillary network in the concrete. This can be achieved by the addition of a filler such as silica fume to the mix or by the addition of long-chain fatty acids.

- **Mineral admixtures:** pozzolans, frequently blended with cements, are a group of compounds that react with lime and water to produce a cementitious material. They increase the chemical resistance of cement paste by reducing the Portlandite content in the cement paste, and in the case of fine admixtures such as silica fume they reduce the permeability of the concrete. Mineral admixtures are often low-cost source materials and therefore reduce the cost of the concrete.

Structure of hardened concrete

For background reading, see references 16–18.

The durability of concrete *vis-à-vis* many aggressive environments is linked to the permeability of the concrete. The voids formed within the cement paste, approximately 15% of the total volume, will consist of discrete voids and interconnecting voids that may form an open capillary structure through the cement paste; such a capillary structure will result in a highly permeable concrete. It can be argued that permeability of concrete is one of the key issues to be considered when assessing the durability of a concrete.

Porosity and permeability are not synonymous terms: a material can be highly porous without being permeable. The difference between the two terms hinges on the interconnection between the voids in the material: a permeable material must have a network of interconnecting channels. When considering durability the pore structure of the material is an important issue. Deterioration is often caused by the transport of aggressive agents into the concrete matrix; therefore, limiting this process improves durability. Three transport mechanisms are considered in concrete technology:[19]

- diffusion
- permeation
- capillary flow.

These mechanisms are characterized by transport parameters. Permeability specifically refers to steady saturated flow under pressure; however, it is often used in a more generic sense.

Diffusion

Diffusion is the mechanism whereby a gas or chemical species moves through the concrete by virtue of a concentration gradient. In the case of the diffusion of a gas through concrete the flow rate is influenced by the ionic form of the gas, the pore structure of the concrete and the degree of saturation of the concrete, e.g. water vapour diffusion can take place up to about 70% relative humidity and above that diffusion ceases. In the case of a saturated or very humid concrete, ionic species such as chloride ions may diffuse. The rate of transport in this case is dependent on the diffusion characteristics of the ion species in fluid as well as the material properties. The diffusion of a specific agent through concrete is characterized by the diffusion coefficient D from Fick's law.[20] It should be appreciated that when considering the diffusion of chemical species into concrete there is a possibility of a chemical reaction with the concrete matrix. The migration of chloride or sulfate ions into concrete will chemically and structurally modify the pore structure, thereby causing a variation in the diffusion coefficient.[21]

Permeation

Permeation consists of the transport of a gas or fluid through a medium by virtue of a pressure gradient. In the case of concrete, permeation will occur in a saturated or semi-saturated media. In all of these cases the transport characteristics will be determined by the concrete structure and fluid viscosity. Probably the simplest case of permeation is the flow of water through a saturated sample. In this case the flow can be described by Darcy's law. The coefficient of permeability K characterizes an aspect of concrete durability.

Capillary flow

A concrete can be thought of as a network of fine tubes and, therefore, if the material is in contact with water there will be a capillary effect, i.e. water will be drawn into the concrete by a suction effect. The flow mechanism in this case is often complex: fluid is transported into the concrete where evaporation may occur. In this case, the transport rates are determined by the concrete structure and its degree of saturation, the controlling fluid properties being viscosity and surface tension.

As in the case of diffusion, during permeation and capillary flow, there is always the possibility of the transported fluid provoking a change in the concrete structure. This means that the transport characteristics are time variable.

Factors affecting transport parameters

Void structure

The cement paste can be thought of as having a structure of four components: the hydrated cement products, capillary pores, gel pores and unhydrated cement. The voids in the structure can be broadly classified by their size:

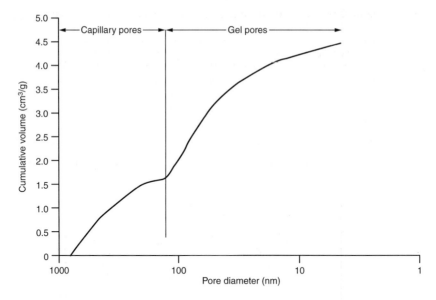

Figure 2.4 *An illustration of the pore size distribution for a cement paste with a high water/cement ratio. Based on Mehta and Manmohan.*[22]

- **capillary pores** $> 0.5\,\mu$m: these pores constitute the spaces between the hydrated cement grains that have not been filled by hydration products. They influence the permeability and ion and gas diffusion through the matrix
- **gel pores** $< 100\,$nm: these pores are the micro-porosity within the CSH structure. They will affect the shrinkage and creep of the hardened cement paste.

A distribution of the size of the pores can be measured by mercury porosimetry, a technique in which mercury is injected into the void structure. The volume injected is noted as the pressure is increased.[23] A typical distribution curve is shown in Figure 2.4.

Two factors have a profound effect on the structure in the cement paste; the water/cement (w/c) ratio and the conditions of cure.

Water/cement ratio
The effect of the water content in the initial mix, characterized by the w/c ratio, dictates the time required to block interconnecting pores and thereby seal the material against external aggression. Table 2.8 gives an indication of the time required to block the capillaries; it is evident that a concrete with a high w/c ratio will be susceptible to ingress of external agents of aggression because of the open pore structure in the cement paste.

Table 2.8 *Relationship between the water/cement ratio and the time taken to disrupt the capillaries.*[24]

Water/cement ratio	Time to block the pore structure
0.40	3 days
0.45	7 days
0.50	14 days
0.55	6 months
0.60	1 year
0.70	Never

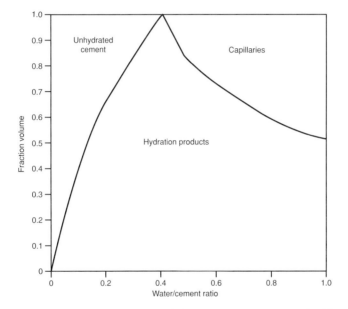

Figure 2.5 *Volumes of the component of the cement paste expressed with respect to the water/cement ratio.*

At the end of this period of hydration it is possible to envisage a distribution of the structure that would be found in a cement paste (Figure 2.5). This figure illustrates the fact that the volume of capillary pore increases progressively above a w/c ratio of about 0.4. It is also of interest to note that the quantity of unhydrated cement increases below 0.4.

To illustrate further the influence of the w/c ratio, the phenomenon of carbonation will be used. Carbonation is a measure of the depth to which the Portlandite in the cement paste is chemically converted to calcium carbonate by carbon dioxide gas; it therefore gives a reasonable insight into the diffusion of CO_2 gas into a concrete. Being a gas diffusion process, it also explains why carbonation is almost non-existent in very humid concretes.

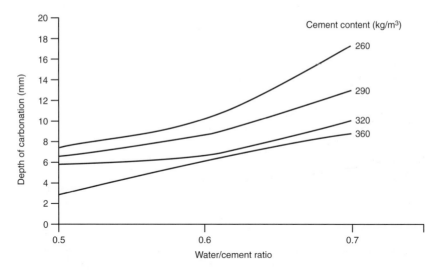

Figure 2.6 *Relationship between depth of carbonation, cement content and water/ cement ratio.*

Figure 2.6 shows the relationship between the depth of carbonation in a concrete at an age of 5 years and the cement content and w/c ratio.[25] It can be seen that the depth of carbonation increases with increased w/c ratio and decreases with increased cement content.

Cure

Good curing conditions ensure that sufficient water is available within the cement paste to allow the hydration reactions to reach completion. If the young concrete is subjected to drying conditions (heat and wind, surface evaporation will draw water from the cement paste), the hydration will cease, resulting in two complementary conditions. Firstly, the filling of the void structure will cease, resulting in an open-pore structure, i.e. high porosity and permeability. Secondly, the hydration process may cease, leaving unhydrated cement throughout the cement paste. The condition of cure will vary through a concrete member. At the centre of a large con- crete mass almost perfect curing conditions will exist; however, the surface of the concrete is highly susceptible to the cure conditions because of its exposure to the external environment and unfortunately this surface skin is, in many cases, the defence against aggressive environments.

Figure 2.7 illustrates the way in which duration of cure affects depth of carbonation after 2 years for three standard cement mortars. The reduction in the duration of cure leads to a mortar that is more susceptible to car- bonation, i.e. the gas diffusivity of the mortar increases with poor cure conditions. The cement with a pozzolan, the Portland fly ash cement, is more susceptible to a poor curing condition; this is explained by the slow

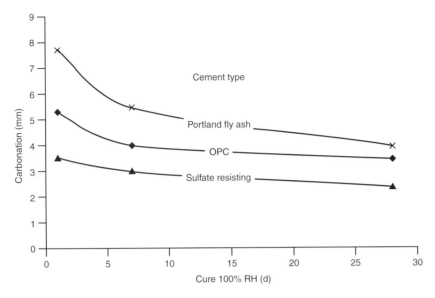

Figure 2.7 *Depth of carbonation after 2 years for three standard cement mortars. Although pozzolans have the effect of reducing water and gas permeability, in this case the depth of carbonation is greater for the Portland fly ash mortar. This anomaly is explained by the low Portlandite content in the mortar owing to its consumption by the pozzolan.*

conversion of the pozzolan to CSH. Poor cure of blended cements will lead to a weaker porous cement paste caused by the early cessation of the Portlandite pozzolan reaction.

Durability and permeability
Many other characteristics of concrete closely correlate to the transport parameters of concrete, such as flexural and compressive strength and abrasion resistance.[26,27] There is also good correlation between the trans-port parameters;[28] however, moisture content has a strong influence on diffusion coefficients and any of the mechanisms involving partially satu-rated concrete, e.g. capillarity.

These characteristics, to a greater or lesser extent, control the durability of a concrete. It can therefore be seen that the pore structure of concrete is a fundamental factor in the study of durability. In principle, a low w/c ratio and good conditions of cure will produce a dense, durable concrete. It should also be remembered that the use of mineral admixtures, notably finely ground material such as silica fume, act to enhance the filling of pores and therefore reduce permeability and thus improve durability. Similarly, it has been shown that the addition of fly ash to OPC has a bene-ficial effect on reducing the diffusion of chloride ions into concrete.

Cement–aggregate interface

The interface between the cement paste and the aggregate is termed the interfacial transition zone (ITZ) or 'aureole de transition'. This constitutes a zone of about 30 μm around an aggregate particle. The impact of this zone on the properties of concrete has been extensively studied,[29] the principal effects being the influence on strength and on permeability. In the case of non-porous aggregates it is considered that a shell of Portlandite and ettringite forms at the interface, the Portlandite crystal axis being perpendicular to the aggregate. It is also considered that these zones have a high porosity. Considering this hypothesis leads one to the conclusion that if these zones overlap, which is a highly probable situation, there will exist porous pathways into the matrix of the concrete. The other consequence of this situation is the existence in a matrix of low-strength paths that may permit the propagation of cracking.

A recent paper by Diamond and Huang[30] used scanning electron microscope (SEM) techniques to examine the nature of the ITZ. They found that portions of the aggregate surface were covered by non-porous deposits of calcium hydroxide and the CSH content within the ITZ was similar to that in the cement paste, and generally concluded that the ITZ had a limited effect on the mechanical properties of the concrete.

Whether the ITZ has significant effects or not, it must be noted that the effect of internal bleeding in a concrete will give rise to voids forming at the aggregate–cement paste interface, this being particularly notable on the underside of flat aggregates.

Most aggregates are porous to some extent. This porosity allows moisture movement between the cement paste and the aggregate, which has a potential effect on the initial quality of the concrete and the long-term performance. In the case of both fresh and hardened concrete a means of water exchange exists between the porous aggregate and the surrounding cement paste. Cracking may also form at the interface, as discussed below.

Segregation and skin effects

The aggregate cement paste structure of a concrete is not isotropic, which is especially true at the concrete surfaces. The evolution of main deleterious effects is controlled by the surface of the material; therefore, there is often a direct link between the surface structure and durability. Variations in the distribution of aggregates will occur within the mass of the concrete, the two principal reasons for this being segregation and skin effects. In fresh concrete the aggregate is effectively in suspension in a viscose fluid, and vibration of the concrete can induce a sedimentation effect, resulting in a gradation of particle sizes over the depth of a section. This effect is most notable at the surface, where a zone develops which is virtually devoid of aggregate. Excessive vibration exacerbates the situation by

Top of the sample

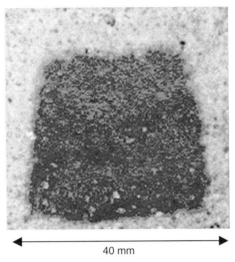

40 mm

Figure 2.8 *Variation in depth of carbonation due to segregation in mortar. The samples have been exposed to air for 2 years and a significant degree of carbonation has taken place. The depth of carbonation is revealed by treating a surface with a chemical indicator (phenolphthalein). The dark area at the centre is sound concrete. It can be seen that the top surface of the concrete is the most susceptible to attack.*

inducing bleeding, thereby increasing the water content at the surface. The development of high porosity due to the bleeding and the high w/c ratio results in a weakened layer of material.

Figure 2.8 shows the depth of carbonation for an exposed mortar. The light area at the perimeter is carbonated. The upper surface of the mortar has the greatest depth of carbonation, which is a direct indication of the effect of segregation that occurs at the time of casting.

A similar non-isotropic effect takes place against the formed face of a concrete section. The variation in concrete structure is frequently termed as the 'skin or surface effect'.[31–35] This skin has been described as a layer that is rich in cement with a low aggregate content. Behind this layer is a zone in which the material consistency is similar to that of a mortar. From these observations one can easily see that this anisotropy of a concrete near the surface can be a factor in the durability of concrete and it should also be borne in mind that these surface zones are the most prone to poor conditions of cure. In conclusion, the surface of concrete can act as a protective shield if properly cured and not prematurely exposed to aggressive elements; however, if the converse is true the outer shield of the concrete will afford no protection and possibly accelerate the process of deterioration.

Internal cracking

All concrete will have some degree of cracking. Freshly hardened concrete has to undergo two processes that may induce internal cracking. Once the structural element is exposed to the ambient environment it begins to cool and suffers a loss of water due to evaporation. The elevated temperatures in the concrete are due to the heat of hydration and, on cooling, thermal shrinkage will occur. The loss of water will result in drying shrinkage of the concrete. Both of these shrinkage mechanisms can, if the concrete is restrained, cause cracking.

In the case of highly porous aggregates the initial degree of aggregate saturation has an immediate impact on the effective w/c ratio of the mix. During the early age of the concrete, water absorption of the aggregate may result in high levels of shrinkage in the area of the aggregate. The induced stresses in the hardening concrete will result in the formation of cracks; this type of effect is highly probable in the case of certain light-weight and recycled aggregates.[36] The possibility of moisture movement between cement paste and aggregate also permits the possibility of shrinkage of the aggregate itself. Aggregate shrinkage often takes the form of peripheral cracking along aggregate boundaries or internal cracking along cleavage planes in the aggregate.[37]

Where high hydration temperatures are allowed to form there will also be a tendency for cracking at the cement paste–aggregate boundary due to differential thermal movement. Thermal cracking is mostly associated with large concrete members and drying shrinkage with slender concrete sections. The internal cracking caused at this stage has a profound effect on the permeability and therefore the durability of the concrete. These problems are very acute when using high-performance concretes. High cement and low water cement contents can achieve high strengths very rapidly, but they tend to have a high modulus of elasticity and low creep. It is for this reason that high cement content cements are very prone to micro-cracking and therefore poor durability performance.

References

1 Mehta P.K. and Monteiro P.J.M. *Concrete structure, properties and materials.* Prentice Hall, Englewood Cliffs, NJ, 1993.
2 Neville A.M. *Properties of concrete*, Longman Higher Education, London, 1995.
3 Pomeroy C.D. Concrete, an overview. In *Construction materials handbook*, ed. D.K. Doran. Butterworth-Heinemann, Oxford, 1994, Ch. 14, pp. 14/1–14/18.
4 Baron J. and Sauterey R. *Le béton hydraulique connaissance et pratique.* Presses Ponts et Chausées, Paris, 1982, p. 558.
5 *Lea's chemistry of cement and concrete*, 4th edn, ed. P.C. Hewlett. Arnold, London, 1998, p. 1053.

6   Folliot A. Le ciment. In *Le béton hydraulique*. Presses Ponts et Chausées, Paris, 1982, Ch. 2, pp. 39–58.

7   Taylor H.F.W. *Cement chemistry*, 2nd edn. Thomas Telford, London, 1997.

8   Dron R. and Voinovitch I.A. *L'activation hydraulique des laitiers, pouzzolanes et cendres volantes*. Presses de l'École Nationale des Ponts et Chaussées, Paris, 1982, Ch. 13, pp. 237–245.

9   Pitts J. Concrete aggregates. In *Construction materials handbook*, ed. D.K. Doran. Butterworth-Heinemann, Oxford, 1994, Ch. 16, 16/1–16/22.

10 Young J.F., Mindness S., Gray R.J. and Bentur A. *The science and technology of civil engineering materials*. Prentice Hall, Englewood Cliffs, NJ, 1998, Ch. 10.

11 Lesage R. Les granulats. In *Le béton hydraulique*. Presses de l'École Nationale des Ponts et Chaussées, Paris, 1982, Ch. 2, pp. 39–58.

12 Taylor H.F.W. Hydration of Portland cement. In *Cement chemistry*, 2nd edn, ed. H.F.W. Taylor. Thomas Telford, London, 1997, Ch. 7, pp. 187–225.

13 Jauberthie R., Fejean J. and Lanos Ch. Formation et stabilité de l'ettringite responsable de la durée de prise du ciment Portland. In *RX2001 4ième colloque*, Dec. 2001, Strasbourg, France.

14 Hodgkinson L. Admixtures and polymers. In *Construction materials handbook*, ed. D.K. Doran. Butterworth-Heinemann, Oxford, 1994, Ch. 15, pp. 15/1–14/28.

15 Hewlett P.C. *Cement admixtures: uses and applications*, 2nd edn. Longman Scientific and Technical, 1988.

16 Young J.F., Mindness S., Gray R.J. and Bentur A. *The science and technology of civil engineering materials*. Prentice Hall, Englewood Cliffs, NJ, 1998, Ch. 11.

17 Buil M. and Ollivier J.-P. Conception des bétons: la structure poreuse. In *La durabilité des Bétons*. Presses Ponts et Chausées, Paris, 1982, Ch. 4.

18 Folliot A. and Buil M. La structuration progressive de la pierre de ciment. In *Le béton hydraulique*. Presses de l'École Nationale des Ponts et Chaussées, Paris, 1982, Ch. 14, pp. 247–259.

19 Kropp J. Transport mechanisms and definitions. *Rilem Report 12, Performance criteria for concrete durability*, eds J. Kropp, H.K. Hilsdorf. E&FN Spon, London, 1995, pp. 4–14.

20 Crank, J. *The mathematics of diffusion*. Oxford University Press, London, 1975, p. 414.

21 Hearn N. and Morley C.T. Self sealing properties of concrete. *Materials and Structures* 1997, **30**, 404–411.

22 Mehta P.K. and Manmohan D. Pore size distribution and permeability of hardened cement paste. *7th Congrès International de la Chimie des Ciments*, Paris, 1980, 3 (7–1), pp. 1–11.

23 Diamond S. A critical comparison of mercury porosimetry and capillary condensation pore size distribution of Portland cement pastes. *Cement and Concrete Research* 1971, **1**, 531–546.

24 Powers T.C. Capillary continuity or discontinuity in cement paste. *Journal of the PCA Research and Development Laboratories* 1959, **1**(2), 38–48.

25 Moll H.L. Über die Korrosion von Stahl. *Beton Deutscher Ausschuss für Stahlbeton* 1964, **167**, 23–61.

26 Kropp J. Relationship between transport characteristics and durability. *Rilem Report 12, Performance criteria for concrete durability*, eds J. Kropp, H.K. Hilsdorf. E&FN Spon, London, 1995, pp. 97–137.

27 Hilsdorf H.K. Concrete compressive strength, transport characteristics and durability. *Rilem Report 12, Performance criteria for concrete durability*, eds J. Kropp, H.K. Hilsdorf. E&FN Spon, London, 1995, pp. 165–197.

28 Nilsson L. and Luping T. Relationship between different transport parameters. *Rilem Report 12, Performance criteria for concrete durability*, eds J. Kropp, H.K. Hilsdorf. E&FN Spon, London, 1995, pp. 15–32.

29 Maso J.-C. La liaison pâte-granulats. In *Le béton hydraulique*. Presses de l'École Nationale des Ponts et Chaussées, Paris, 1982, Ch. 14, pp. 247–259.

30 Diamond S. and Huang J. The ITZ in concrete – a different view based on image analysis and SEM observations. *Cement and Concrete Composites* 2001, **23**, 179–188.

31 Emerson M. Mechanisms of water absorption by concrete. In *Protection of concrete*, eds R.K. Dhir, J.W. Green. E&FN Spon, London, 1990, pp. 689–700.

32 Harwood P.C. Surface coating specification criteria. Deterioration and repair of reinforced concrete in the Arabian Gulf. *Proc. of 3rd Int. Conf.*, Vol. 1, Bahrain Society of Engineers, Bahrain, 1989, pp. 635–646.

33 Dhir R.K., Hewlett P.C. and Chan Y.N. Near surface characteristics of concrete abrasion resistance. *Materials and Structures* 1991, **24**, 122–128.

34 Rendell F., Jauberthie R. and Camps J.-P. The influence of surface absorption on sulfate attack. *Creating with concrete*, Dundee, 6–10 Sept. 1999.

35 McCarter W.J., Emerson M. and Ezirim H. Properties of concrete in the cover zone: developments in monitoring techniques. *Magazine of Concrete Research*, 1995, **47** (172), 243–251.

36 Mesbah H.A., Buyle-Bodin F. and Acker P. Early age shrinkage, mechanical behaviour and cracking of fibre reinforced mortar containing recycled aggregates. *Concrete Science and Engineering* 2000, **2** (June), 71–77.

37 Building Research Establishment. Shrinkage of natural aggregates in concrete. *BRE Digest 357*, Watford, UK, 1991.

Chapter 3

Deterioration of concrete

This chapter aims to give a general background to the deterioration of concrete. Many of the issues raised will be developed further in following chapters.

Broadly speaking, problems with concrete can be classified into four sections:

- **Initial design errors**: either structural or in the assessment of environmental exposure.
- **Built-in problems:** the concrete itself can have built-in problems. A good example of this is alkali silica reaction (ASR), where the alkalinity in the cement paste reacts with the aggregate; this reaction produces an expansive gel that can cause extensive cracking to concrete.
- **Construction defects**: poor workmanship and site practice can create points of weakness in concrete that may cause an acceleration in the long-term deterioration of the structure. A common defect of this type is poor curing of the concrete. This results in a permeable concrete, which can lead to the accelerated deterioration of a structure faced with an aggressive environment.
- **Environmental deterioration:** a structure has to satisfy the requirement of resistance against the external environment. Problems may occur in the form of physical agents such as abrasion, and biological or chemical attack such as sulfate attack from ground water. The key issue here is: what is the mechanism of attack and how severe is it?

Built-in problems in concrete

In its simplest form the principal components of the concrete are cement, aggregate and water. The chemical reaction between the cement and the water produces a stone-like material, not dissimilar to a limestone conglomerate. Steel reinforcement is cast into concrete to augment shear and tensile properties of the material. This steel is generally in a passive condition, i.e.

it will not rust. The following section highlights some of the potential problems that can occur as a result of the materials themselves.

Cement

- **Ordinary Portland cement (OPC):** The significance of the alkali content in relation to possible ASR needs to be considered. Have appropriate precautions been taken to avoid thermal cracking in high cement content deep-section members? Has the concrete been adequately cured and appropriate attention been given to the avoidance (as far as possible) of shrinkage cracking?
- **Sulfate-resisting cement (SRC):** Where chloride salts are present, sulfate-resisting cement may show an increased susceptibility to reinforced corrosion. The C_3A plays a role in binding the migrating chloride ions.
- **High alumina cement (HAC):** The converted form of HAC may show a reduction in strength and become more susceptible to certain forms of chemical attack. More recently, there has been increased concern over the occurrence of carbonation in HAC which has been responsible for problems with reinforcement corrosion in addition to the problems due to loss of strength as a result of conversion.

Steel

The reinforcement may consist of one or more of a variety of types: mild steel, high yield steel, weathering steel, ferritic, stainless or galvanized steels.

The steel itself is rarely responsible for problems with reinforced concrete; problems more frequently encountered are caused by corrosion and positioning of the reinforcement. Steel in sound concrete, which has a pH of about 13, is in a passive state, that is, a protective chemical skin forms over the surface of the reinforcement that will inhibit corrosion. If this skin breaks down, corrosion will commence. Coated bars are sometimes used where corrosion problems are envisaged. Early experience with this type of reinforcement showed that damage to the coating due to poor site handling could be responsible for accelerated corrosion damage.

Chemical admixtures

- **Retarders:** older structures, pre-1970s, in which calcium chloride was used may be suffering from reinforcement corrosion.
- **Plasticizers:** a very large overdose can retard the setting of the concrete. If super-plasticizers are used with air-entraining admixtures they may reduce the entrained air content.
- **Freeze–thaw resistance:** an overdose of an air-entraining agent will produce a reduction in strength and may aggravate frost damage.

Insufficient admixture may not be effective in preventing frost damage.
- **Mineral admixtures:** the potential problems encountered with pozzolans include a slow gain in early age strength and incomplete hydration of the pozzolans.

Water

A good general rule is that if the water is drinkable it should be suitable for concreting purposes, and the vast majority of water employed for concrete will have been drawn from the mains. However, circumstances may dictate the use of other sources and (very occasionally) problems may be encountered. Consideration should be given to the presence of dissolved salts, organic matter, sugar, lead, etc., which can retard the setting of the cement. Much more frequently, the effects of an excessive or inadequate water content will be encountered, resulting respectively in a porous concrete or a concrete which is difficult to compact.

Aggregates

The aggregates used in Europe reflect the vast range of different geological types of material. Typical problems encountered with aggregates are set out in Tables 3.1 and 3.2.

It is sometimes necessary to use local aggregates that are in many respects unsatisfactory. An example of a complex aggregate cement interaction is shown in Figure 3.1. The case discussed refers to the use of laterites, a decomposed granitic material, for road construction in Senegal. The stabilization of the material can be achieved with either lime or cement. In this case a reactive aggregate is used positively because of its pozzolanic properties.

Construction defects

Poor site practice during the fabrication of concrete work may lead to problems that will resurface possibly years after the structure is put into service. A description of defects in concrete emanating from the construction phase can be found in Department of Energy[1] and Ministère de l'Équipement.[2]

- **Poor compaction** of concrete and cold joints is usually readily apparent in a structure. Such areas provide porous areas of concrete that are open to aggressive environmental attack. Poor site control can lead to high water content in the concrete, often resulting from operatives increasing the workability by adding water to the mix before placing; this will result in low strength and highly porous concrete.

Table 3.1 *Potentially harmful physical characteristics of aggregates due to mechanical properties*

	Effect
Particle shape	Elongated particles with an aspect ratio (length/width ratio) greater than 3 will lead to a low workability, poor compaction and high void content, if either insufficient water is added or a plasticizer is not used
Badly and gap-graded aggregates	A badly graded aggregate will affect the workability, compaction and consequently the strength
Mica	Disruption to the characteristics of fresh concrete: high water content needed to achieve workability. Causes poor bonding.
Clay coating to the aggregate	May cause poor aggregate bonding and possibly aggregate shrinkage
Porous aggregates	Disruption to the characteristics of fresh and early-age concrete followed by the possibility of aggregate shrinkage in the hardened concrete. They will also be very susceptible to freeze–thaw action
Mechanically weak aggregates, poor physical properties such as aggregate crushing value or flakiness	Poor bonding and deterioration of the aggregate due to environmental conditions

Table 3.2 *Potential chemical problems associated with aggregates*

	Effect	Examples
Organic matter	Disruption to setting and hydration. Staining of the hardened concrete, potential for expansion	Minerals containing coal or lignite
Sulfates	Promote the formation of expansive ettringite	Gypsum
Sulfides	Particle weakness and formation of expansive compounds. Can stain concrete	Pyrite contamination
Chlorides	Can contribute to the depassivation of reinforcement	Poorly washed sea-dredged aggregates
Alkali-reactive aggregates	Formation of an expansive gel	Flints and cherts
Carbonate aggregates	Possible formation of thaumasite. Alkali carbonate reaction	Dolomitic limestone
	Dolomitic limestone can also lead to the release of magnesium into the cement paste which forms the non-cementitious brucite and hydrated magnesium silicates	

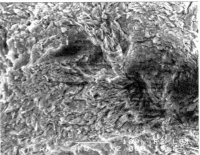

Figure 3.1 *Scanning electron micrographs of Feldspar found in laterite. The image to the left shows the deteriorated nature of the feldspar (×150) This deterioration is due to kaolinization, a process in which the feldspar undergoes an alteration to amorphous SiO_2 and kaolin (a clay). The amorphous SiO_2 is pozzolanic in the presence of lime. However, the particles are covered in with a thin film of kaolin (right-hand micrograph ×2000) which can prevent the pozzolanic reaction and ultimately create a plane of weakness between the quartz and the cement paste.*

- **Poor curing** of a concrete is one of the most common causes of loss of material durability. The cure of concrete is intrinsically linked to the hydration of the cement and therefore the strength gain and the porosity of the material.
- **Thermal cracking** of concrete can occur in large pours. Typically, concrete can gain in temperature about 14°C per 100 kg of cement in a cubic metre of concrete. In large pours this sets up a thermal gradient, with the outer part of the concrete cooling more rapidly than the core. This puts the outer skin in tension and small cracks form. With the addition of subsequent drying shrinkage, the cracks can become quite large. This effect is compounded by the fact that the coefficients of expansion of aggregate and cement paste are different.

Plastic settlement and shrinkage

During the setting of the concrete the material undergoes a transition from a liquid to a solid, during which it is possible for cracking to take place. Plastic cracking generally appears within the first few hours after placing and is of two distinct types.

- **Plastic settlement cracks** are typically found in columns, deep beams or walls. The problem tends to occur in concretes with a high water/cement (w/c) ratio which have suffered from bleeding. The concrete 'hangs up' on the steel, slumping between it, with cracks forming over the line of the steel (Figure 3.2). Caught early enough, re-vibration of the concrete can repair the damage while the concrete is still plastic. Figure 3.2 shows a core taken through a plastic

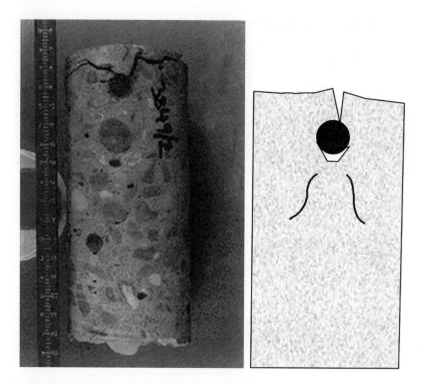

Figure 3.2 *Example of plastic settlement: the concrete separation over the steel can be seen. Cracking resulting from plastic settlement forms within hours of casting and runs along the line of the reinforcement.*

settlement crack in a car park. The reinforcement is corroded owing to the creation of an easy pathway for chlorides.

- **Plastic shrinkage** results when the rate of water loss from evaporation exceeds the rate at which bleeding (movement of water to the surface) occurs. This leads to a surface network of cracks as one would see in a clay that has dried in the sun. The result of this is the formation of shrinkage cracks in the concrete which establish pathways that will open the material to external attack. Not surprisingly, it is more of a problem in hot, dry climates, but can easily occur in flat slabs cast on hot days, especially where inadequate attention has been paid to protection and curing.

Note that neither of these forms of cracking should be confused with drying shrinkage cracks, which only occur after a considerable time.

Low cover to the reinforcement is a common site defect. This will result from either poor concrete design detailing or poor site control. The concrete between the surface and the steel acts as a protective layer that prevents the reinforcement from rusting. If this cover is reduced the time taken to initiate the corrosion of the reinforcement is reduced. This will manifest itself in cracking of the concrete and rust staining at an early age. An example of corrosion damage to a structure with low cover is shown in Figure 3.5.

Environmental deterioration

This section reviews some of the reasons for the deterioration of concrete that occurs during the service life of the structure. The flow diagram after Hermann[3] (Figure 3.3), attempts to assemble some of the main causes of concrete deterioration. However, such attempts to rationalize the deterioration process are intrinsically flawed because of the complexity of the material and the exposure environment. In many cases a concrete will suffer from several concomitant vectors of attack. Any mechanism of attack that weakens the materials will generally open the way for other mechanisms to weaken the material further.

Physical actions

The physical processes that cause a deterioration of concrete can be broadly classified as surface and internal damage. The most common cause of surface damage is abrasion, normally resulting from the action of water- or wind-borne solids. The internal damage to concrete results from the formation and propagation of micro-cracking due to stress levels in the material. The right-hand side of Figure 3.3 indicates some of the principal physical actions causing concrete deterioration:

- excessive induced stress
- abrasion
- fire damage
- internal stresses due to mechanical processes, corrosion of reinforcement and freeze–thaw damage.

Externally induced stress

Actual structural failure, or even structural cracking, is only rarely encountered, but it is important to differentiate between cracking for structural reasons and other causes. Such an assessment should only be carried out by a structural engineer, but an initial inspection by a materials engineer may highlight other (much more likely) causes of cracking. The materials engineer will know when the advice of a structural engineer is required. Figure 3.4 illustrates structural cracks at a column–beam connection.

As a concrete or mortar is stressed by either internal or externally applied loads, micro-cracking will occur within the material. The micro-crack structure within the concrete matrix is very much dependent on the level of loading, which will influence its permeability and creep characteristics.

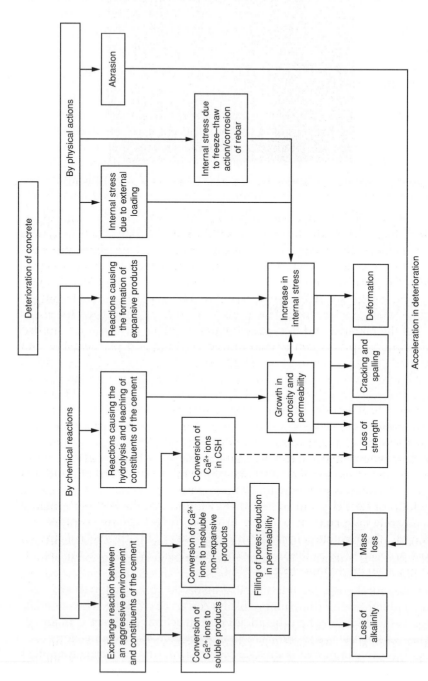

Figure 3.3 *Possible mechanisms involved in the deterioration of concrete.*

Figure 3.4 *Structural cracking in a joint between a column and a cross-beam.*

Variations in the cement paste properties occur at the interface with aggregates, thus causing discontinuities within the material which become a nucleus for crack propagation.[4] At a stress below 30% of failure stress the situation is stable; between 30 and 50% the micro-cracks increase in length, width and number; above 50% the cracks bridge between grains of aggregate and, finally, above 70% there is a spontaneous crack growth leading to a continuous crack system. This process of cracking may open the structure of the material and thus raise the permeability. Gérard *et al.*[5] studied the relationship between stress-induced cracking and permeability, and reported that increases in permeability of the order of 10^2 to 10^3 were noted as the stress level was raised from 40 to 70% of the failure strength of the concrete. It was also noted that there was a trend for cracks of less than 0.1 mm to heal, whereas larger cracks remained as permanent defects.

The mechanism. The work of Piasta and Schneider[6] considered the changes in mechanical properties of concrete under sustained compressive load when subjected to sulfate attack. They concluded that the level of compressive stress significantly affected the rate of deterioration, and that high stress and severe chemical attack conditions reduced the residual modulus of elasticity and strength. They found that the best performance of concrete occurred at a stress level of 20–35% of the compressive strength. The reduction in material performance at high levels of stress is most probably due to the increased permeability, resulting from the formation of an interconnecting crack system.

Abrasion

Water often carries suspended solids, frequently sands and gravels; the transport of such materials will cause abrasion to the surface of the concrete. In sewers this effect can be controlled to some degree by limiting the flow velocities; however, modern sewer-cleaning methods using high-pressure jetting will result in very high abrasive forces to virtually all parts of the sewer.[7]

Abrasion of a concrete surface will have two major influences on the durability of concrete: first, the removal of the outer cement-rich skin of the concrete, and secondly, the physical removal of sound and degraded material, opening the pore structure to further attack. In cases where abrasion does occur, the damaged and therefore irregular face of the material presents a higher surface area liable to attack. The abrasion resistance of a concrete is closely linked to its strength, that is, the strength of the cement paste and the aggregates.

Internal stresses

If stress levels can be induced within the concrete matrix that exceed its tensile stress, micro-cracking will occur. This will lead to an opening of the matrix of the material, resulting in an acceleration of the process of deterioration.

Corrosion of reinforcement.[8]   Steel reinforcement is normally chemically protected from corrosion by the alkaline nature of the concrete. Steel embedded in sound concrete at a pH of the order of 12–13 will have a natural protection against corrosion: the steel is in a **passive state**. If this alkalinity is lost through carbonation or if chlorides are present which can break down this immunity, then corrosion can occur. This loss of the protective layer is known as **depassivation**. An example of the type of damage associated with reinforcement corrosion is shown in Figure 3.5. When cover is low, the onset of corrosion will occur at an earlier age.

If corrosion occurs in the reinforcement in an environment where oxygen is available, the corrosion product (ferrous oxide, red rust) will form. This product will occupy a greater volume than the steel and the resulting expansion will cause high levels of internal stress within the concrete. If no oxygen is available, for example in a submerged structure, the ferric corrosion product is non-expansive and no cracking will occur. The two common causes of depassivation are:

- carbonation: often occurs in urban areas where there are high levels of carbon dioxide
- chloride damage: generally found in marine conditions or on road structures where salt is used for de-icing.

Carbonation.   Carbonation-induced corrosion tends to affect large areas of the bar, causing a gradual loss of section over a relatively wide area. The corrosion problem is obvious before serious damage can be done because the concrete cover will spall.

The process of carbonation requires the presence of water and carbon dioxide gas in the pore structure; consequently, carbonation will not occur in saturated concrete. The worst condition for carbonation is found at a relative humidity (RH) between 50 and 60%. The carbon dioxide reacts with water in the pores and capillaries to form carbonic acid, which in

Figure 3.5 *Example of corrosion damage to a column. The damage usually takes the form of spalling or delamination of the concrete cover. Initially, cracks form along the line of the steel and as the corrosion progresses the cover concrete is detached. In this case, the steel cage is misplaced due to lack of spacers.*

turn reacts with the Portlandite and calcium silicate to form calcium carbonate.

$$CO_2 + H_2O \rightarrow H_2CO_3$$
$$H_2CO_3 + CaO \cdot SiO_2 \cdot H_2O \rightarrow CaCO_3 + SiO_2 + 2H_2O$$

and

$$H_2CO_3 + Ca(OH)_2 \rightarrow CaCO_3 + 2H_2O$$

These reactions are associated with a 12% increase in volume;[9] this has the effect of reducing the average pore size and thereby reducing further infiltration. The carbonation reaction is also associated with a reduction in pH of the concrete from 12 to about 8. If the carbonation reaches the reinforcement, depassivation of the steel will take place and a corrosion process will ensue, if sufficient oxygen and moisture are present. In an aerobic environment the corrosion products of steel are expansive and this will lead to large-scale cracking of the concrete.

The rate at which carbonation occurs is a function of concrete quality, mainly the w/c ratio and the compaction. It is generally accepted that the rate of the carbonation reaction is inversely proportional to the square root of the age of the structure. If the depth of carbonation is taken in millimetres and the age of the structure in years, the constant of proportionality is approximately unity.

So, for K (rate constant) = 1,

$$\text{Rate of carbonation (mm/year)} = \frac{1}{(\text{Age in years})^{0.5}}$$

(NB. The rate applies *only* at the particular age chosen. The rate cannot be used for other ages.)

$$\text{Depth of carbonation (mm)} = 2 \times (\text{Age in years})^{0.5}$$

Recent research[10] suggests that the square root relationship holds only at about 50% RH. At higher humidity the power function drops off, so that above 90% RH the depth of carbonation is likely to equate to the (age in years)$^{0.3}$ and continues to fall at higher humidity. The effect of this is that the carbonation depth will be lower for concrete continuously exposed to higher humidity.

On this basis, even with a cover of only 10 mm, steel reinforcement should be safe for more than 25 years. In practice, however, carbonation often occurs more rapidly, either because the concrete is excessively permeable or due to micro-cracking in the concrete providing secondary paths to the steel other than by normal diffusion processes. As described in Chapter 2, excessive permeability can result from a high w/c ratio and from poor curing of the cover concrete; these effects will have a significant effect on the rate of carbonation. Most modern specifications fail to recognize the importance of curing on concrete quality.

Chlorides (marine structures and those in contact with de-icing salts). With chlorides, a different mechanism often occurs, causing very localized severe loss of section. This can occur without disruption of the cover concrete and almost total loss of section can occur before problems become apparent at the surface. Where pre-stressed steel is used, catastrophic failures have occurred with no prior warning on structures which had been load tested shortly before the failure occurred. Chloride-induced damage to concrete has been widely reported.[11,12] As in the case of carbonation, this process requires the presence of pore water. Chlorides will be present either as an 'Internal' chloride, i.e. chloride added to the concrete at the time of mixing, or as an 'External' chloride, i.e. chloride ingressing into the post-hardened concrete. Some of the internal chloride binds with the

C_3A, forming calcium chloroaluminate. In this form, the chloride is insoluble in the pore fluid and is not available to take part in damaging corrosion reactions. When the concrete carbonates (calcium carbonate formation due to the reaction of Portlandite with CO_2) the bound chlorides are released. In effect, this provides a higher concentration of soluble chloride immediately in front of the carbonation zone. The external chlorides migrate into the concrete by diffusion and absorption; the chloride ions diffuse through the pore water, so the permeability of the concrete is important in this process. Once the chloride concentration at the reinforcement reaches approximately 0.4% by weight of cement, depassivation and corrosion of the steel occur (this limit can also be expressed in terms of the ratio Cl^-/OH^-; in excess of 0.6 depassivation occurs). The normal form of the corrosion takes place in an aerobic environment and leads to expansive corrosion products that cause spalling of the concrete. In cases where depassivation takes place in an anaerobic environment, it is possible to set up a macro cell,[13] the cathodic reaction taking place in an aerated zone of the structure and the anodic (corrosion) process occurring at the depassivated area. The corrosion product formed is magnetite and thus is non-expansive. This effect has been found to be responsible for the corrosion of pre-stressed cables in bridge structures where complete grouting of the cable ducts has not occurred.

Salt damage

This problem manifests itself with a surface scaling similar to that found with freeze–thaw damage. The problem occurs where there is a high salt concentration in the ground water; sodium sulfate and sodium carbonate have been signalled as being particularly damaging.[14] Capillary action takes the salt-charged ground water above ground level. As the pore water evaporates the salt concentration in the pores becomes sufficiently high to generate high crystallization pressures (Figure 3.6). Changes in relative humidity (RH) and temperature provoke cycles of dissolution and recrystallization, and each time the crystallization causes internal pressure within the pore structure close to the surface.

Freeze–thaw action

Freeze–thaw action occurs owing to the formation of ice crystals within the matrix of the concrete. On freezing there is an increase in volume of the lens of water and this leads to the generation of internal stress. The effect becomes progressively more severe as the matrix of the material is fissured: the opening of the structure permits the formation of a greater number of sites at which the effect can take place. The damage takes the form of cracking and surface scaling. This kind of attack often occurs in tandem with other mechanisms.

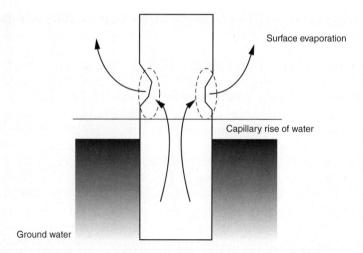

Figure 3.6 *Salt damage to concrete. The ground water rich in salts is drawn above the standing water level by capillary action. The areas of damage occur above ground level where the surface is subject to evaporation. The damage normally takes the form of surface scaling.*

Aggregate shrinkage

It is often assumed that the aggregate is inert in the concrete; however, this is not always the case. Some, mostly igneous, aggregates can contain inclusions of weathered material in the form of clay minerals. These minerals, in common with the clays encountered in the ground, swell in the presence of moisture and shrink as they dry out. They can cause excessive drying shrinkage of the concrete and can cause a random crack pattern not unlike that encountered with ASR (Figure 3.7).

Fire damage

Three principal types of alteration are usually responsible.

- **Cracking and micro-cracking in the surface zone:** this is usually sub-parallel to the external surface and leads to flaking and breaking away of surface layers. Cracks also commonly develop along aggregate surfaces, presumably reflecting the differences in coefficient of linear expansion between cement paste and aggregate. Larger cracks can occur, particularly where reinforcement is affected by the increase in temperature.
- **Alteration of the phases in aggregate and paste:** the main changes occurring in aggregate and paste relate to oxidation and dehydration. Loss of moisture can be rapid and probably influences crack development. The paste generally changes colour and various colour zones can develop (Figure 3.8). A change from buff or cream to pink tends to occur at about 300°C and from pink to whitish grey at about 600°C. Certain types of aggregate also show these colour

Figure 3.7 *Severe example of cracking due to aggregate shrinkage; it is accompanied by pop-outs where frost has attacked the porous aggregate.*

Figure 3.8 *Example of fire-damaged concrete. The surface to the right has been exposed to the heat and there is a change in colour to red.*

changes, which can sometimes be seen within individual aggregate particles. The change from a normal to light paste colour to pink is most marked. This occurs in some limestones and some siliceous rocks, particularly certain flints and chert; this can also be found in the feldspars of some granites and in various other rock types. It is likely that the temperature at which the colour changes occur varies somewhat from concrete to concrete, and if accurate temperature profiles are required some calibrating experiments need to be carried out.

- **Dehydration of the cement hydrates:** this can take place within the concrete at temperatures a little above 100°C. It is often possible to detect a broad zone of slightly porous light buff paste which represents the dehydrated zone between 100 and 300°C. It can be important, in reinforced or pre-stressed concrete, to establish the maximum depth of the 100°C isotherm.

Chemical and biochemical attack

These mechanisms of attack are summarized on the left-hand side of Figure 3.3. The severity of the attack will depend on a variety of factors, such as the concentration of a species, time of exposure and temperature. Similarly, the resistance of the concrete to attack will depend on the type of cement, w/c ratio, type of curing, etc. As discussed in Chapter 2, some of the key parameters controlling durability are the transport parameters of the concrete and its strength. The rate of diffusion and permeation controls the rate of ingress of the aggressive agent and the strength controls the formation of micro-cracking.

For background reading, see references 9, 15 and 16.

Conversion of Ca^{2+} ions to soluble products

This group of reactions result in the decalcification of the concrete, that is a progressive opening of the structure and a reduction in its pH. The opening of the concrete matrix results in a higher porosity and therefore the material is prone to other forms of attack. In addition, the surface of the material is mechanically weakened and finally the reduction in pH leads to a depassivation and consequently corrosion of the reinforcement.

Ammonium salts (chemical storage and wastewater systems). The reaction between certain ammonium salts and concrete has long been recognized as potentially aggressive; ammonium chloride, phosphate, sulfate, sulfide, sulfite and bicarbonate are considered the most harmful, whereas ammonium carbonate, oxalate and fluoride are harmless.[15] The principal process involved in the deterioration of concrete in the presence of ammonium compounds involves the reaction between the Portlandite (lime) and the ammonium salt:

$$Ca(OH)_2 + 2NH_4Y \rightarrow CaY_2 + 2H_2O + 2NH_3$$

where Y is the cation associated with NH_4^+.

Ammonium sulfate and ammonium nitrate are usually considered to be the most aggressive of the ammonium salts. Ammonium nitrate provokes a solubilization and leaching of the lime in the cement paste. This is accompanied by a reduction in the pH of the concrete, thus creating a danger of depassivation of the steel in reinforced concrete. The leaching of the lime also leads to a weakening of the matrix of the material with some cracking.[17]

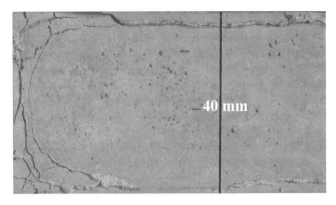

Figure 3.9 *Damage to an ordinary Portland cement mortar due to ammonium sulfate solution. The solution causes a liberation of ammonium gas and a high degree of swelling in the mortar. This swelling (1.0 mm/m) and washing with a water jet provoked the crack damage seen here.*

Ammonium sulfate is the most destructive of the sulfate salts. Concrete immersed in a solution of the sulfate undergoes considerable swelling and cracking, which is the principal reason for its loss of strength (Figure 3.9). The depth of decalcification is not as great as that seen in the nitrate; however, the displaced calcium forms a surface deposit of gypsum.[18,19]

Organic compounds (chemical storage and wastewater treatment). Acetic and formic acids are by-products of biodegradation. The acids react with the Portlandite, $Ca(OH)_2$, and the products of hydration in the cement paste. Bajza *et al.*[20] proposed the following reactions:

$2CH_3COOH + Ca(OH)_2 \rightarrow (CH_3COO)_2Ca + 2H_2O$
  Acetic acid   Portlandite

$6CH_3COOH + 3CaO \cdot 2SiO_2 \cdot 3H_2O \rightarrow 3(CH_3COO)_2Ca + 2SiO_2 \cdot aq + nH_2O$
  Acetic acid   CSH

$6CH_3COOH + 3CaO \cdot Al_2O_3 \cdot 6H_2O \rightarrow 3(CH_3COO)_2Ca + Al_2O_3 \cdot aq + nH_2O$
  Acetic acid C_3A hydrate

The SiO_2 is insoluble, whereas the Al_2O_3 aqueous hydrated oxide is in a solid state at pH above 3.5. Other compounds known to attack concrete include beer, milk and vegetable oils.

The effect of sugar solution on concrete was studied by Skenderovic *et al.*[21] The change in chemical composition and pore structure was studied by X-ray diffraction and mercury intrusion porosimetry. It was found that the rate of attack increased up to a solution concentration of 10% and then became constant. There was a significant increase in the pore volume: 45% in all pores and 64% in open pores. The analysis indicated a leaching of calcium, manganese, iron and silicon.

Microbial weathering processes (wastewater systems and atmospheric pollution). Many investigators, e.g. Berthelin[22] and Silva,[23] have reported on the microbial weathering of rocks and soils. This solubilization is also seen in concrete where it is exposed to humid conditions and organic growth becomes established. The process involves the chemical and mineralogical modification of different materials and essentially involves the solubilization of mineral elements (Si, Al, Fe, Mn, Mg, K and Na) from silicates, oxides, phosphates, carbonates and sulfides.

An example of a bacterial reaction is often found in wastewater systems. In this case hydrogen sulfide gas is oxidized into sulfuric acid by a bacterial reaction; this oxidization occurs within an organic 'slime' layer by aerobic *Thiobacillus* bacteria. The end-product of this reaction is sulfuric acid at pH 2, which is capable of solubilizing the Portlandite in the cement paste.[24–27]

Conversion of Ca^{2+} ions to insoluble non-expansive products

It is possible for an external environmental condition to produce a reaction with concrete that has the effect of sealing the surface porosity of a concrete. A well-known example of this phenomenon is carbonation. The reaction between carbon dioxide and Portlandite produces calcium carbonate. The carbonate is very slightly expansive and has the effect of filling the capillary structure of the concrete and thus reducing its porosity. It has long been accepted that a little carbonation has a beneficial effect.

In the case of concrete exposed to sea water the presence of magnesium chloride (2000 mg/l) is a common form of attack on the Portlandite through the base exchange reaction:

$$Ca(OH)_2 + MgCl_2 = CaCl_2 + Mg(OH)_2$$

Magnesium hydroxide (Brucite) is a soft, paste-like deposit that is not easily leached out of the pore structure, and to a certain extent protects the concrete. In extreme cases where the concrete is of poor quality, or where sea water enters cracks in the concrete in the intertidal zone, it is possible to find a total degeneration of the cement paste.

Another example of the formation of a protective layer is seen in the case of sulfuric acid attack. In this case, a dense layer of gypsum forms on the surface of the concrete, thus inhibiting further attack. Unfortunately, this seemingly promising method of improving durability fails in a real environment, as chemical attack from a solution is nearly always combined with surface shear and abrasion associated with the movement of the liquid. The surface layer of gypsum is weak in shear and therefore easily removed, thus exposing new surfaces to attack.

Reactions causing the formation of expansive products

Sulfate attack. Concrete buried in soils or ground water containing high levels of sulfate salts, particularly in the form of sodium, potassium or magnesium salts, may be subjected to sulfate attack under damp conditions. As sulfates migrate into a concrete a reaction occurs between these sulfates and calcium monosulfate to form calcium sulfoaluminate (ettringite). This reaction is expansive and the tensile forces developed cause cracking in the cement matrix. Past experience has shown that true sulfate attack is rare in concrete, only occurring with very low cement content concretes, with less than about $300 \, kg/m^3$ of cement. As a guide, levels of sulfate above about 4% of cement (expressed as SO_3) may indicate the possibility of sulfate attack, provided that sufficient moisture is present. Sulfate attack requires prolonged exposure to damp conditions. Sulfate attack on concrete has been well documented; the widely accepted view was that deterioration was due to the formation of ettringite, a highly expansive crystal form:

For $pH > 12.5$

$$2(3CaO \cdot Al_2O_3 \cdot 12H_2O) + 3(X_2SO_4 \cdot nH_2O) \rightarrow 3CaO \cdot Al_2O_3 \cdot 3CaSO_4 \cdot 31H_2O$$

$$C_3A \qquad\qquad\qquad\qquad\qquad\qquad Ettringite$$

$$+ \; 2Al(OH)_3 + 6X \cdot OH$$

$$+ \; (3n - 13)H_2O$$

Wang[28] argued that much of the sulfate damage is due to the degeneration of the $Ca(OH)_2$ in the outer layers of the hydrated cement. In these areas the pH will be lower and ettringite cannot exist:

$$Ca(OH)_2 + X_2 \cdot SO_4 \cdot nH_2O \rightarrow CaSO_4 \cdot 2H_2O + 2X \cdot OH + (n - 2)H_2O$$

$$Portlandite \qquad\qquad\qquad\qquad Gypsum$$

If the hydroxide products are highly soluble, X ammonium for example, the hydroxide will leach out; the cement paste is decalcified and the deterioration process accelerated.

Thaumasite formation. Thaumasite was recognized as a deterioration product in the 1960s. Recently there have been several cases of deterioration of foundations in the 1990s which have been attributed to the formation of thaumasite,[29] $CaSiO_3 \cdot CaCO_3 \cdot CaSO_4 \cdot 15H_2O$. This form of attack occurs in cold, wet conditions and is caused by the reaction between sulfates in ground water and carbonates in the aggregate of the concrete. It is reported that dolomitic limestone aggregates are most susceptible. In theory, an external source of carbonates could also cause this reaction.

The combination of these factors can cause an unusual reaction between the cement, the lime and the sulfate, to form **thaumasite**, a sulfate mineral. The effect is to cause serious damage and softening of the exposed outer surface of the concrete (assuming an external source of sulfate). Figure 3.10 shows one of the affected foundations and also a damaged road in north Texas which showed serious heaving of the road surface over a gypsum-rich soil when a lime soil stabilizer was used under the roadway. It should be noted that sulfate-resisting cement has not proved to be any more resistant than normal Portland cement in resisting this type of attack.

Figure 3.10 *Examples of thaumasite attack in concrete. Foundation damaged by thaumasite attack (left) and a road in north Texas heaving as a result of thaumasite/ettringite attack (right).*

Delayed ettringite formation (DEF).[30] As Don Hobbs, formerly of the British Cement Association, said: 'DEF is not as simple as ABC'!

During the hydration process, it is quite normal for ettringite (calcium sulfoaluminate) to form inside the concrete. This mineral is normally associated with sulfate attack, but in the context of a setting concrete is quite normal. Any expansion resulting from its formation is taken up in the still plastic concrete. However, if the temperature of the concrete exceeds about 60°C then formation of the ettringite can be delayed until after the concrete has hardened. In this situation, if a source of moisture is present, then the concrete can suffer from quite severe expansion and cracking. There is a correlation with the alkali content of the cement: the higher the alkali content, the lower the temperature at which DEF can occur.

DEF produces a map cracking on the surface. Internally, ettringite forms a shell around the aggregate. As the ettringite expands it produces cracks which in turn fill with ettringite. These effects are illustrated in Figure 3.11.

Aggregate reaction. Deterioration of concrete can be caused by an internal reaction, alkali aggregate reaction (AAR), within the matrix of the hardened concrete.

In the 1930s a number of cases of severe cracking in structures were noted in the USA. AAR was identified on Jersey at the Val de la Mare dam in 1971 and in subsequent years numerous further cases of aggregate reaction were observed in the UK. The extent of damage in some cases was so extensive that the structures had to be demolished. The crack damage was found to exist at depth within the concrete and was not a surface-controlled phenomenon as seen in previous cases. The reaction responsible

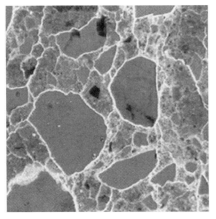

Figure 3.11 *Delayed ettringite reaction. Ettringite formation around the aggregate (left) and a detail of ettringite filling of a crack (right).*

for the damage was identified as AAR: a reaction between the aggregate and a high alkali content in the pore water of the concrete.[31,32]

The principal source of high alkali is the cement (high potassium and sodium contents in the cement) which leads to high concentration of KOH and NaOH in the pore water; this can elevate the pH of the pore water to the order of 13.5–13.9 (normally 12.7–13.1). Two principal types of reaction can exist:

- **alkali carbonate reaction** occurs principally in dolomitic limestones. The high concentrations of NaOH and KOH lead to the formation of brucite and calcite. The crystallization pressure induced by the formation of these products leads to cracking in the concrete matrix;
- **alkali silica reaction** is the most common form of the problem and is associated with the **reactive** aggregates. It is believed that the most expansive behaviour is associated with poorly organized silica forms (opal, chert, cryptocrystalline quartz), whereas quartz shows no noticeable reactivity. One of the most frequently found aggregates in affected concrete is chert. This is a common constituent of many gravel aggregates, but a number of other geological types may be reactive, such as strained quartz in sands and some quartzites. Some Irish aggregates, notably greywackes, have been found to be susceptible to ASR. The reaction causes the formation of a hydroscopic gel which, when under humid conditions, can swell: the resulting pressures have been reported of the order of 5–6 MPa and are therefore capable of causing tensile failure of the aggregate and the cement matrix. The reaction product is a hygroscopic gel that takes up water and swells. This may create internal stresses sufficient

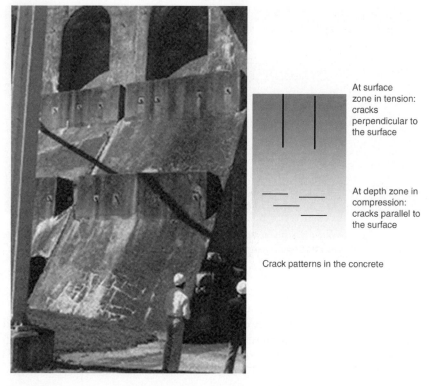

At surface zone in tension: cracks perpendicular to the surface

At depth zone in compression: cracks parallel to the surface

Crack patterns in the concrete

Figure 3.12 *Cracking to the Beauharnois Dam on the St. Lawrence Seaway. The expansion caused by the alkali silica reaction on this structure caused it to grow several centimetres in length and distorted the turbines providing hydroelectric power for the region.*

to crack the concrete. These reactions tend to be quite slow reacting and damage can take 20–50 years to become serious.

The cracking of the concrete is triggered by internal expansion of the gel that puts the concrete under compression. If the concrete is retrained, by reinforcement for example, cracking will occur parallel to the main restraint (along the line of the principal reinforcement). A free unrestrained surface, as seen in the photograph in Figure 3.12, will exhibit a map or Isle of Man cracking. In extreme cases this will be associated with gel leaching out of the cracks.

Interactions

Working through the preceding sections it should have been apparent that envisaging deterioration as a linear process is entirely misleading. As one

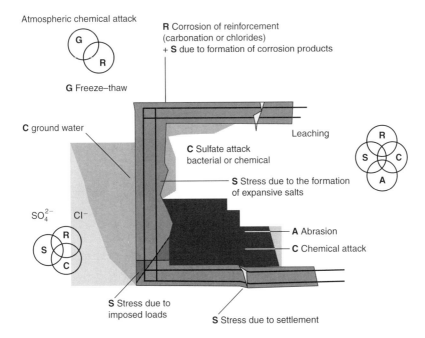

Figure 3.13 *Possible interactions in the mechanisms of attack that can be exerted on a structure. The interactions are represented by Venn diagrams.*

vector of attack breaks down the defences of concrete another vector of attack will surely follow. A common theme in this 'wheel of misfortune' is often the permeability of the concrete matrix. As the concrete surface is weakened, its permeability to liquid and gas is increased and the result is evident, acceleration of attack. Figure 3.13 illustrates the possible combinations of attack that may occur in a structure. While reading the following chapters one should continually think of the importance of the interactions between vectors of attack, as failure to do so will give a simplified and often erroneous picture of the situation that is evolving.

Stress/chemical interaction

Nägele introduced an innovative approach to the analysis of the effect of stress plus chemical corrosion;[33,34] the method of analysis proposed was based on the work of brittle fracture of ceramics. This work examined the relationship of the probability of material failure [expressed as 'time to failure' (τ)] to the level of applied stress and the nature of the aggressive environment. Figure 3.14 illustrates the influence of sustained load on the ultimate strength of the material. The value β represents the strength properties of the material, with β_0 being the initial strength; as the sustained load on the material is increased, the life of the material is reduced.

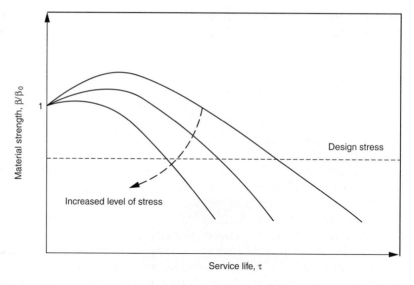

Figure 3.14 *Relationship between design life and level of stress for a material subjected to stress and an external environmental aggression.*

Concrete *vis-à-vis* an aggressive environment

In conclusion to this chapter it is worth stressing the link between material selection and durability. Two issues are paramount: the permeability of the concrete and the physicochemical stability of the concrete. Durability and permeability of concrete are in the majority of cases closely linked. A dense, well-designed mix normally yields strong, impermeable material that will resist an aggressive environment. In the case of reinforced concrete a good depth of cover has the virtue of delaying the time of the onset of corrosion damage. However, in making these generalizations one should also bear in mind the dangers from within: high-performance concretes may suffer from internal cracking due to thermal cracking and reactive aggregates are a constant cause of concern in material selection.

Material selection

Chemical resistance of cements

It will be noted throughout this brief review that many of the forms of chemical attack involve a reaction with the Portlandite in the cement matrix. Several indices have been produced to predict the resistance of Portland cement to chemical attack; many of these indices link chemical resistance to the Portlandite content in the cement paste. An example of this type of index is the aggressivity modulus (Grün, 1944), which predicts the sulfate resistance of cements:

$$\text{Aggressivity modulus} = \frac{SiO_2 + Fe_2O_3}{CaO + MgO + Al_2O_3}$$

This modulus implies that the higher the value the higher the resistance. The modulus favours low lime, magnesia and alumina contents. In general, Portland cements lie in the range 0.25–0.40; blast furnace cements will generally have values in excess of 0.4.

For many years it has also been recognized that high C_3A content potentially leads to the formation of ettringite. To specify concrete resistant to sulfate attack the C_3A content is normally limited to 3.5%.

It should be noted that current practice tends to the use of blended cements rather than SRC. Blended cements containing pozzolans lead to a consumption of Portlandite in the pozzolanic reaction, thereby increasing the chemical resistance of the material. The strength and permeability of such cement pastes show an inferior performance at a young age; however, as the pozzolans react to form CSH the ultimate performance characteristics of the concrete become superior to those of the Portland cement. The Building Research Establishment[35,36] has conducted an extensive study into the effectiveness of OPC PFA mixes in sulfate and marine conditions. It was generally concluded from this study that the blended PFA Portland cements showed an improved sulfate-resisting performance even when the C_3A content of the cement was as high as 14%. Many European countries recommended pozzolanic cements for improving resistance against acid conditions; however, this effect is probably marginal.[37,38]

Resistance against ASR
Certain aggregates are highly susceptible to this type of reaction and regulation for the selection of aggregates is covered by several norms (e.g. EN 206 NF P 19-001). The alkalinity of the cement is a critical factor in the reaction; the limiting alkalinity is of the order of 0.6% equivalent Na_2O. For any given alkali content there is often a proportion of reactive aggregate that produces a maximum expansion, this is known as the pessimum proportion. Pozzolans are recognized as playing a role in the reduction in the risk of ASR; this is possibly due to a consumption of the Portlandite leading to lower gel swelling and finally the ultimate formation of a dense pore structure that inhibits ion migration. It has also been suggested that the high proportion of highly reactive silica in the pozzolan moves the proportion of reactive aggregate in the concrete to a value above the pessimum proportion.

Resistance against corrosion of reinforcement

Chlorides. The selection of cement type to minimize the damage induced in marine conditions can create a problem, as the sulfate content in sea water would dictate a low C_3A cement. However, it is widely reported that the hydrated C_3A and C_4AF compounds bind chloride ions and therefore reduce the rate of chloride ion penetration. This contradictory demand has led to many codes and norms specifying low w/c ratio cements (low permeability) and high cover for marine environments. (This has the effect of delaying the time required for chloride ions to reach the steel and therefore cause depassivation.) The addition of pozzolans to a cement paste has the effect of decreasing the diffusion coefficient of chloride ions; however, there will be an associated reduction in the OH^- concentration in the cement paste, this being most notable in the case of silica fume blends. As depassivation occurs when the Cl^-/OH^- exceeds 0.6 it can be seen that pozzolans potentially have a detrimental effect on resistance against chlorides. However if a concrete with OP pozzolan blended cement is properly cured it may show improved resistance to chloride penetration.[36,39]

Carbonation. Initial surface carbonation has a pore-blocking effect and consequently reduces the diffusion of CO_2 into the concrete. In general it has been noted that the carbonation rate increases with the amount of pozzolan in a blended cement. This is particularly true in the case of slags; at 50% replacement the carbonation is about 1.5 times that found with a comparable Portland cement. In the case of PFA replacement, concretes with and without replacement show similar carbonation rates for replacements up to 30%. At 50% PFA replacement, carbonation rates are significantly higher.[40] As with all concretes containing pozzolans, the regime of cure is vital: a prolonged period of cure will greatly improve the permeability of the material against external attack and therefore improve carbonation resistance. Another way to improve carbonation resistance is by the use of calcareous filler in the mix, which has the effect of reducing permeability.[41]

References

1 Department of Energy. *The classification and identification of typical blemishes on the surface of concrete underwater*. OTH-84-206. H.M. Stationery Office, London, 1984.

2 Ministère de l'Équipement. *Défauts d'aspect des pavements en béton. Guide technique*. Ministère de l'Équipement du logement, des transports et de la mer, 1991.

3 Hermann K. Substances which have a chemical action on concrete. *Bulletin du Ciment* 1995, **63** (11), 3–11.

4 Maso J.-C. La liaison pâte-ciment. In *Le béton hydraulique*. Presses Ponts et Chaussées, Paris, 1982, pp. 247–259.

5 Gérard B., Breysse D., Ammouche A., Houdusse O. and Didry O. Cracking and permeability of concrete in tension. *Materials and Structures* 1996, **29** (April), 141–151.

6   Piasta W.G. and Schneider U. Deformation and elastic modulus of concrete under sustained compression and sulfate attack. *Cement and Concrete Research* 1992, **22**, 149–158.

7   Cant J. and Trew J. High-pressure water jetting: avoiding damage to sewers. *Journal of the Institution of Water and Environmental Management* 1998, **12** (4), 265–267.

8   Pullar-Strecker P. *Corrosion damaged concrete – assessment and repair.* Butterworth, London, 1987.

9   Moorhead D.R. Factors affecting the durability of concrete. *Concrete 95, Towards better concrete structures*, Vol. 1, Brisbane, Australia, 1995, pp. 287–293.

10 Parrott L.J. Carbonation induced corrosion. Proc. of a one day seminar, *Improving Civil Engineering Structures – Old and New.* Geological Society, London, 31 January 1995. Geotechnical Publishing, 1994.

11 Delagrave A., Pigeon M. and Revertegat E. Influence of chloride ions and pH on the durability of high performance cement pastes. *Cement and Concrete Research* 1994, **24** (8), 1433–1443.

12 Tuutti K. Effect of cement type and different additions on service life. In *Concrete 2000*, eds R.K. Dhir, M. Roderick Jones. E&FN Spon, London, 1993, pp. 1285–1295.

13 Rendell F. and Miller W. Macro-cell corrosion of reinforced concrete in marine structures. In *Corrosion of reinforcement in concrete*, eds C.L. Page, K. Treadaway, P. Bamforth, Society of Chemical Industry. Elsevier Applied Science, London, 1990, pp. 167–177.

14 Haynes H., O'Neil R. and Mehta P.K. Concrete deterioration from physical attack by salts. *Concrete International* 1996 (January), 63–68.

15 Biczok J. *Concrete corrosion and concrete protection*, 8th edn. Akademiai Jiadl, Budapest, 1972.

16 Lea F.M. *The chemistry of cement and concrete*, 3rd edn. Edward Arnold, London, 1970.

17 Ukraincik V., Bjegovic D. and Djurekovic A. Concrete corrosion in a nitrogen fertiliser plant. In Durability of building materials and components, *Proc. 1st Int. Conf.*, Ottawa, Canada, 1978, pp. 30–41.

18 Rendell F. and Jauberthie R. The deterioration of mortar in sulfate environments. *Construction and Building Materials* 1999, **13**, 321–327.

19 Rendell F., Jauberthie R. and Camps J.P. The effect of surface gypsum deposits on the durability of cementitious mortars under sulfate attack. Concrete Science and Engineering. *RILEM 2000*, **2** (April), 231–244.

20 Bajza A., Rouseciva I. and Uncik S. Corrosion of hardened cement paste by acetic and formic acid. *Slovak Journal of Civil Engineering* 1994, **4** (2), 26–34.

21 Skenderovic R., Oponcky L. and Franc L. Study of the mechanism and dynamics of concrete corrosion in sugar solution. Paper 7, *Durability of building materials and compounds*, eds J.M. Baker, P.J. Nikon, A.J. Majumdar, J.F. Clifton. E&FN Spon, Brighton, 1991, pp. 65–96.

22 Berthelin J. Microbial weathering process. In *Microbial geochemistry*, ed. W.E. Krumbien. Blackwell Scientific, Oxford, 1983, pp. 223–262.

23 Silva M.R. The influence of the proximity of the sea on the degradation of concrete by micro organisms. *Proc. 5th Int. Conf. on Structural Faults and Repairs*, ed. M.C. Forde, Vol. 2. Edinburgh Technical Press, Edinburgh, 1993, pp. 21–29.

24 Aziz M.A. and Koe L.C.C. Durability of concrete sewers in aggressive sub-soils and groundwater conditions. In *Geotechnical aspects of restoration works*, eds A.S. Balasubramniam *et al.* Balkema, Rotterdam, 1990, pp. 299–310.

25 Forrester J.A. An unusual example of concrete corrosion induced by sulfate reducing bacteria in sewers. Cement and Concrete Association, London, *Technical Report RA 320*, 1959.

26 Fjerdingstad E. Bacterial corrosion of concrete in wastewater. *Water Pollution Research* 1979, **3**, 21–30.

27 Bos P. and Kumen J.G. Microbiology of sulphur oxidising bacteria. *Proc. Conf. Micro. Corr.*, National Phys. Lab., Teddington, UK, 1983, pp. 8–10.

28 Wang J.G. Sulfate attack on hardened cement paste. *Cement and Concrete Research* 1994, **24** (4), 735–743.

29 Crammond N.J. and Halliwell M.A. Assessment of the conditions required for thaumasite form of sulfate attack. In *Mechanisms of chemical degradation of cement based systems*, eds K.L. Scrivener and J.F. Young. Mat. Res. Soc., Boston, 1996.

30 Lawrence C.D. Delayed ettringite formation: an issue? In *Materials science of concrete*, eds J. Skalny, S. Mindness, American Ceramic Society, Westervuill, 1995, pp. 113–154.

31 Hobbs D.W. *Alkali-silica reaction in concrete*. Thomas Telford, London, 1988.

32 ACI. State of the art report on alkali-aggregate reactivity. ACI Committee 221, Dec. 1998.

33 Nägele E. *Spannungskorrosion zementgebundener Baustoffe in Ammonium-salzlösungen Habilitationsschrift*. Universität Gsamthochschule Kassel, 1990.

34 Nägele E. New and powerful method for the evaluation of multi-parameter corrosion tests. *Cement and Concrete Research* 1995, **25** (6), 1209–1217.

35 Matthews J.D. Performance of pfa concrete in aggressive conditions. 1: Sulfate Resistance. Building Research Establishment Laboratory Report, Building Research Establishment, UK, 1995.

36 Matthews J.D. Performance of pfa concrete in aggressive conditions. 2: Marine conditions. Building Research Establishment Laboratory Report, Building Research Establishment, UK, 1995.

37 Eglinton M.S. *Review of concrete behaviour in acidic soils and ground waters*. CIRIA Technical Note 69.

38 Futtuhi N.I. and Hughes B.P. Effect of acid attack on concrete with different admixtures and protective coatings. *Cement and Concrete Research* 1983, **13**, 655–665.

39 Thomas M.D.A. and Matthews J.D. Chloride penetration and reinforcement corrosion in fly ash concrete exposed to a marine environment. In *Third CANMENT/ACI International Conference on Concrete in Marine Environment*, New Brunswick, Canada, 1996, pp. 317–338.

40 Thomas M.D.A. and Matthews J.D. Carbonation of fly ash concrete. *Magazine of Concrete Research* 1992, **44** (160, Sept.), 217–228.

41 Ranc R. and Cariou B. Quality and durability of concretes made with fillerized cements, the French experience. *BRE Seminar on Limestone-filled Cements*, London, 1989.

Chapter 4

In situ investigation of concrete deterioration

For background reading, see references 1–5.

The assessment of a structure will be driven by either a routine inspection requirement, a change in use of the structure or a potential problem being identified by the owner/operator. The preliminary study will involve an assessment of the condition of a structure, in general terms a preparation of a statement of fact: what is found to exist. This study will lead to the diagnosis of the problem, an estimation of service life and, if appropriate, the proposal of a remedial action plan. This chapter sets out some of the techniques that are used during the first stages of the assessment of concrete condition, i.e. the *in situ* study of the structure.

Before starting the *in situ* testing it is recommended that as much information as possible is assembled.

- What design approach was adopted and what assumptions were made concerning the environmental exposure?
- What materials, cement and aggregate types were used? Were cement replacements or blended cements used, etc.?
- What were the conditions at the time of construction: time of year, temperatures, etc.?
- What is the history of the structure? Has there been overloading or fire damage, etc.?

Experience has shown that a number of testing methods are of proven value in determining the extent of deterioration of a concrete structure and in identifying those areas where remedial measures are necessary. While the list of tests given below is not exhaustive, it does include most of the common tests as well as one or two lesser known techniques.

The two-stage approach

Any investigation can conveniently be split into two stages:

- **stage 1:** an initial survey to identify the cause of the problems

- **stage 2:** an extension of the stage 1 survey, perhaps using a limited number of techniques to identify the extent of the defects revealed by stage 1.

The advantages of such an approach are clear. In the stage 1 survey, work can be carried out on selected areas showing typical defects but choosing these, as far as possible, from areas with simple access, i.e. ground level, roof level, from balconies, etc. Occasionally, a lightweight scaffold tower or an electrically powered hydraulic lift can be used to advantage. One or more areas apparently free from defect would also be examined in this initial survey as it is frequently found that, by comparing good areas with bad, the reason for the problems emerges by simple comparison.

In stage 2, once the defects have been identified, it is often necessary to quantify the extent of the problems. This may be as simple as carrying out a cover meter survey over the whole structure, where low cover has been identified as the problem, to the application of one or more of the other techniques described below.

Visual surveys

After collecting as much background information as possible, any testing problem should begin with a thorough visual survey of the structure. This may conveniently be recorded on a developed elevation, paying particular attention to the following defects:

- **construction defects**
  - honeycombing due to poor compaction or grout loss
  - areas in which remedial finishing work had already been carried out
- **in-service defects**
  - external contamination or surface deposits
  - wet or damp surfaces
  - varying colour or texture
  - hollow surfaces
  - cracks or crazing
  - spalling
  - corrosion of steel and rust staining.

Throughout the course of any investigation colour photographs should be taken of points of particular interest.

Important information concerning the condition of a structure can be gleaned from crack and spalling damage. Cracks as small as 0.05 mm can be seen on a dry, clean surface; however, the cracks become more discernible at crack widths greater than 0.1 mm. Crack widths can be measured with a simple crack gauge; when it is necessary to monitor crack movement a demountable strain gauge (Demec gauge) can be used. At the inspection stage careful logging of crack patterns is important, but a background 'feel' for what is the root of the problem enables a more efficient search for information. An interpretation of crack type and cause will be discussed later in the book. The reader is reminded of the concluding section of Chapter 3 which warned of the complexities resulting from the interaction of several vectors of attack. Figure 4.1 sets out a description of typical surface cracking

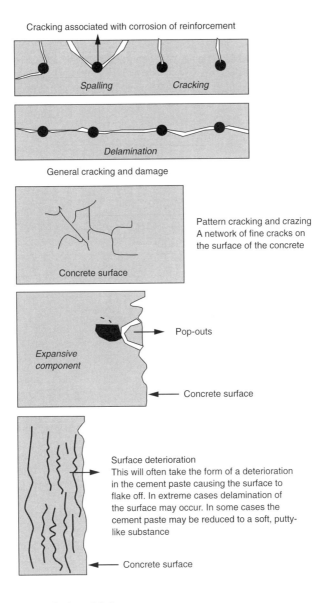

Figure 4.1 *Description of defects.*

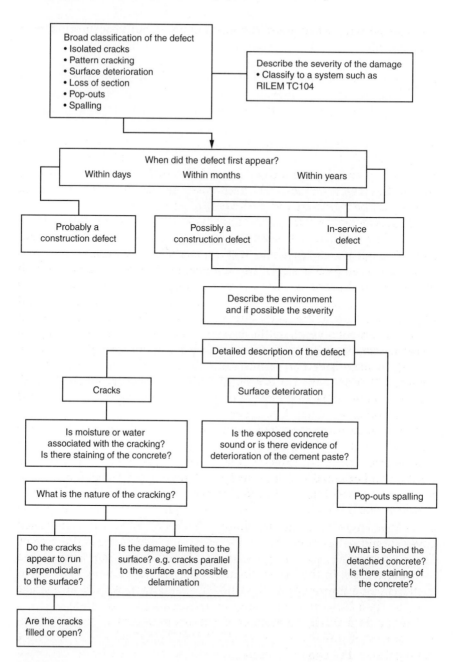

Figure 4.2 *Describing defects (RILEM TC104[6]).*

and damage, while a suggested flow diagram for visual inspection is shown in Figure 4.2.

Cover

Covermeter survey

Adequate cover to the steel reinforcement in a structure is important to ensure that the steel is maintained at a sufficient depth into the concrete, so as to be well away from the effects of carbonation or from aggressive chemicals. However, excessively deep cover has its own problems: crack widths may be increased and the lever-arm decreased.

All covermeters are electromagnetic in operation. Electric currents in a coil winding in the search head generate a magnetic field, which propagates through the concrete and will interact with any buried metal present, such as reinforcing steel. The interaction will be due to either or both of two physical properties of the steel: its magnetic permeability and its electrical conductivity. The interaction causes a secondary magnetic field to propagate back to the head where it is detected by a second coil or, in some instruments, modifies the primary field. The signal received will increase with increasing bar size and decrease with increasing bar distance (cover). By making certain assumptions about the bar, and specifically by assuming that only one bar is present within the primary magnetic field, the instrument can be calibrated to convert signal strength to distance and hence to indicate the depth of cover.

If there is more than one bar within the range of the primary field the instrument will receive a stronger signal and indicate a shallower cover than the true cover. The skilled operator will always carefully map out the position and orientation of the steel, breaking out some steel if necessary, to ensure that accurate results are obtained.

British Standard 1881: Part 204: 1988[7] requires that when measuring cover to a single bar under laboratory conditions, the error in indicated cover should be no more than $\pm 5\%$ or $\pm 2\,mm$, whichever is the greater. For site conditions, an average accuracy of $\pm 15\%$ or $\pm 5\,mm$ is suggested as being realistic in the British Standard. Recent developments in covermeters are now improving on this, showing better than 8% with an average of better than 2% over a wide range of bar sizes, lapped bars, etc.

The standard also lists a number of extraneous factors, which are potential sources of error. Those concerned with magnetic effects from the aggregates or the concrete matrix, and those due to variations in cross-sectional shape of the bars, should not affect the modern covermeter, but care must always be taken when dealing with multiple bars[8] and the effects of adjacent steel, such as window frames or scaffolding, as mentioned earlier.

Assessing concrete quality

Ultrasonic pulse velocity (UPV) measurement

Theory

The velocity of ultrasonic pulses travelling in a solid material depends on the density and elastic properties of that material; these are often related to their elastic stiffness. The measurement of UPV in such materials can often be used to indicate their quality as well as to determine their elastic properties. In addition to assessment of the material properties, ultrasonic techniques can be used for finding internal defects in the material.

The velocity of a pulse of longitudinal ultrasonic vibrations travelling in an elastic solid is given by the following equation:

$$v = \sqrt{\frac{E}{\rho}} \times \frac{(1 - v)}{(1 + v)(1 - 2v)}$$

where E is the dynamic elastic modulus, ρ is the density and v is Poisson's ratio. This equation may be considered to apply to the transmission of longitudinal pulses through a solid of any shape or size provided the least lateral dimension (i.e. the dimension measured perpendicular to the path travelled by the pulse) is not less than the wavelength of the pulse vibrations. The frequencies suitable for these materials range from about 20 to 250 kHz, with 50 kHz being appropriate for the field testing of concrete. These frequencies correspond to wavelengths ranging from about 200 mm for the lower frequency to about 16 mm at the higher frequency.

Method of testing

For assessing the quality of materials from UPV measurement, this measurement must be of a high order of accuracy. This is done using an apparatus that generates suitable pulses and accurately measures the time of their transmission (i.e. transit time) through the material tested, i.e. between the transmitting and receiving transducers. The distance that the pulses travel in the material (i.e. the path length) must also be measured to enable the velocity to be determined; the path length and transit time should each be measured to an accuracy of about ±1%.

Figure 4.3 shows how the transducers may be arranged on the surface of the specimen tested, the transmission being either direct, indirect or semi-direct. The direct transmission arrangement is the most satisfactory one since the longitudinal pulses leaving the transmitter are propagated mainly in the direction normal to the transducer face. The indirect arrangement is possible because the ultrasonic beam of energy is scattered by discontinuities within the material tested, but the strength of the pulse detected in this case is only about 1 or 2% of that detected for the same path length

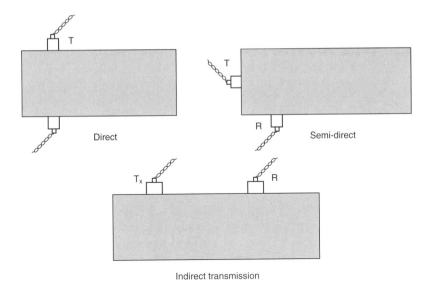

Figure 4.3 *Transmission modes used in ultrasonic testing.*

when the direct transmission arrangement is used. The indirect configuration of transducers is often the only option open when inspecting slabs and walls; the values of velocity are slightly less in this mode than in the direct mode.

Applications
The pulse velocity method of testing may be applied to the testing of plain, reinforced and pre-stressed concrete, whether it is pre-cast or cast *in situ*. The measurement of pulse velocity may be used as follows.

- To determine the **homogeneity of the concrete:** measurement of pulse velocities at points on a regular grid on the surface of a concrete structure provides a reliable method of assessing the homogeneity of the concrete.
- **Detection of defects, large voids or cavities:** when an ultrasonic pulse travelling through concrete meets a concrete–air interface, there is a negligible transmission of energy across this interface, so that any air-filled crack or void lying directly between the transducers will obstruct the direct beam of ultrasound, when the void has a projected area larger than the area of the transducer faces. It is sometimes possible to make use of this effect for locating flaws, etc., but it should be appreciated that small defects often have little or no effect on transmission times.

- **Estimating the depth of surface cracks:** an estimate of the depth of a crack visible at the surface can be obtained by measuring the transit times across the crack for two different arrangements of the transducers placed on the surface.
- **Monitoring changes in concrete with time:** changes occurring in the structure of concrete with time caused either by hydration (which increases strength) or by an aggressive environment, such as frost or sulfates, may be determined by repeated measurements of pulse velocity at different times. Changes in pulse velocity are indicative of changes in strength and measurements can be taken over progressive periods of time on the same test piece or concrete product. This facility is particularly useful for following the hardening process during the first 2 days after casting, and it is sometimes possible to take measurements through formwork before it is removed. This has a useful application for determining when formwork can be removed or when pre-stressing operations can proceed.

Estimation of strength after fire damage

Figure 4.4 shows that a good correlation could be obtained to estimate the residual crushing strength of the concrete after heating from pulse velocity tests.

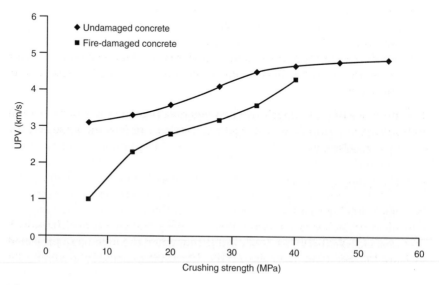

Figure 4.4 *Curves for residual strength with ultrasonic pulse velocity (UPV) for fire-damaged concrete: UPV measurements may be used to assess the extent of damage.*

Estimation of strength
Concrete quality is generally assessed by measuring its cube (or cylinder) crushing strength. There is no simple correlation between cube strength and pulse velocity, but the correlation is affected by:

- type of aggregate
- aggregate/cement ratio
- age of concrete, size and grading of aggregate
- curing conditions.

Further details of the effects of these factors may be found in references 9–12.

In practice, if pulse velocity results are to be expressed as equivalent cube strengths, it is preferable to calibrate the particular concrete used by making a series of test specimens with materials and mix proportions the same as the specified concrete, but having a range of strengths. The pulse velocity is measured for each specimen, which is then tested to failure by crushing. The range of strength may be obtained either by varying the age of the concrete at test or by introducing a range of water/cement (w/c) ratios. The curve relating cube strength to pulse velocity is not likely to be the same for these two methods of varying strength, but the particular method chosen should be appropriate to the test purpose required.

Influence of test conditions
In the interpretation of ultrasound observations it must be kept in mind that the pulse velocity in concrete may be influenced by several factors:

- path length
- lateral dimensions of the specimen tested
- presence of reinforcing steel: steel is an excellent conductor of sound and so will effectively short-circuit the path length
- moisture content of the concrete: water is also an excellent conductor of sound; in saturated concrete it may be impossible to detect cracks and voids.

The influence of path length will be negligible provided it is not less than 100 mm when 20 mm size aggregate is used or not less than 150 mm for 40 mm size aggregate.

Surface hardness methods

Schmidt rebound hammer
One of many factors connected with the quality of concrete is its hardness. The rebound principle for measuring hardness is widely accepted as a method for assessing concrete quality. The most popular equipment, the Schmidt rebound hammer, invented by Ernst Schmidt in the late 1940s, has been in use world-wide for many years. Recommendations for the use of the rebound method are given in BS 1881: Part 202 (45) and ASTM C805 (46).

Rebound test equipment and operation

A spring-controlled hammer mass slides on a plunger within a tubular housing. The plunger retracts against a spring when pressed against the concrete surface and this spring is automatically released when fully tensioned, causing the hammer mass to impact against the concrete through the plunger. When the spring-controlled mass rebounds, it takes with it a rider, which slides along a scale and is visible through a small window in the side of the casing. The rider can be held in position on the scale by depressing the locking button. The equipment is very simple to use and may be operated either horizontally or vertically, and either upwards or downwards.

The plunger is pressed strongly and steadily against the concrete at right angles to its surface, until the spring-loaded mass is triggered from its locked position. After the impact, the scale index is read while the hammer is still in the test position. Alternatively, the locking button can be pressed to enable the reading to be retained or results can automatically be recorded by an attached paper recorder. The scale reading is known as the rebound number and is an arbitrary measure, since it depends on the energy stored in the given spring and on the mass used. This version of the equipment is most commonly used and is most suitable for concrete in the 20–60 N/mm^2 strength range. Electronic digital reading versions of the equipment are available.

Procedure

The equipment is very sensitive to local variations in the concrete, especially to aggregate particles near the surface, and it is therefore necessary to take 12 readings in the area of interest and to average the results obtained. A recommended procedure can be found in BS 1881: Part 202. The surface to be measured should be smooth, clean and dry, but if measurements need to be taken on trowelled surfaces, the surface can be smoothed using the carborundum stone provided with the instrument.

Theory, calibration and interpretation

The test is based on the principle that the rebound of an elastic mass depends on the hardness of the surface upon which it impinges and in this case will provide information about a surface layer of the concrete, defined as no more than 30 mm deep. The results give a measure of the relative hardness of this zone and this cannot be directly related to any other property of the concrete. Energy is lost on impact owing to localized crushing of the concrete and internal friction within the body of the concrete, and it is the latter, which is a function of the elastic properties of the concrete constituents, that makes theoretical evaluation of test results extremely difficult. Many factors influence the results, but all must be considered if the rebound number is to be empirically related to strength.

Factors influencing test results
Results are significantly influenced by all of the following factors:

- **mix characteristics:**
  - cement type
  - cement content
  - coarse aggregate type
- **member characteristics:**
  - mass
  - compaction
  - surface type
  - age, rate of hardening and curing type
  - surface carbonation
  - moisture condition
  - stress state and temperature.

Since each of these factors may affect the readings obtained, any attempts to compare or estimate concrete strength will be valid only if they are all standardized for the concrete under test and for the calibration specimens. These influences have different magnitudes. Hammer orientation will also influence measured values, although correction factors can be used to allow for this effect.

Other near-surface strength tests
These include the Windsor probe, the BRE internal fracture tester, various break-off devices, and the CAPO and Lok tests used extensively in Scandinavia, and increasingly in the United Kingdom.

Radar profiling

Since the early 1990s sub-surface impulse radar has been used increasingly to investigate civil engineering problems and, in particular, concrete structures. Electromagnetic waves, typically in the frequency range 500 MHz to 1 GHz, will propagate through solids, with the speed and attenuation of the signal being influenced by the electrical properties of the solid materials. The dominant physical properties are the electrical permittivity, which determines the signal velocity, and the electrical conductivity, which determines the signal attenuation. Reflections and refractions of the radar wave will occur at interfaces between different materials, and the signal returning to the surface antenna can be interpreted to provide an evaluation of the properties and geometry of sub-surface features.

Radar systems
There are three fundamentally different approaches to using radar to investigate concrete structures.

- **Frequency modulation:** the frequency of the transmitted radar signal is continuously swept between pre-defined limits. The return signal is mixed with the currently transmitted signal to give a difference frequency, depending on the time delay and hence the depth of the reflective interface. This system has seen limited use to date on relatively thin walls.
- **Synthetic pulse radar:** the frequency of the transmitted radar signal is varied over a series of discontinuous steps. The amplitude and phase of the return signal is analysed and a time domain synthetic pulse is produced. This approach has been used to some extent in the field and also in laboratory transmission line studies to determine the electrical properties of concrete at different radar frequencies.
- **Impulse radar:** a series of discrete sinusoidal pulses within a specified broad frequency band is transmitted into the concrete, typically with a repetition rate of 50 kHz. The transmitted signal is often found to comprise three peaks, with a well-defined nominal centre frequency.

Impulse radar systems have gained the greatest acceptance for field use and most commercially obtainable systems are of this type. The power output of the transmitted radar signal is very low, and therefore is the least hazardous of the techniques.

Radar equipment
Impulse radar equipment comprises a pulse generator connected to a transmitting antenna. This is commonly of a bow-tie configuration, which is held in contact with the concrete and produces a divergent beam with a degree of spatial polarization. A centre frequency antenna of 1.5 GHz is often used in the investigation of relatively small concrete elements, up to 500 mm thick, while a 500 MHz antenna may be more appropriate for deeper investigations. However, a lower frequency loses resolution of detail despite the improved penetration.

An alternative to using surface-contact antennae is to use a focused beam horn antennae with an air gap of about 300 mm between the horn and the concrete surface. These systems have been used in the USA and Canada to survey bridge decks from a vehicle moving at speeds of up to 50 km/h, principally to detect corrosion-induced delamination of the concrete slab. Operational details are provided in ASTM D4748.

Structural applications and limitations
In addition to the assessment of concrete bridge decks, radar has been used to detect a variety of features buried within concrete, ranging from reinforcing bars and voids to murder victims. The range of principal reported structural applications is summarized in Table 4.1. Interpretation

Table 4.1 *Structural applications of radar*

Reliability	
Greatest ←	→ Least

Determine major construction features
    Assess element thickness
        Locate reinforcing bars
        Locate moisture
            Locate voids, honeycombing, cracking
            Locate chlorides
               Size reinforcing bars
               Size voids
                    Estimate chloride concentrations
                      Locate reinforcement corrosion

of radar results to identify and evaluate the dimensions of sub-surface features is not always straightforward. The radar picture obtained often does not resemble the form of the embedded features. Circular reflective sections such as metal pipes or reinforcing bars, for example, present a complex hyperbolic pattern, due to the diverging nature of the beam. The use of signal processing can simplify the image, but interpretation is still complex. Evaluating the depth of a feature of interest necessitates a foreknowledge of the speed at which radar waves will travel through concrete. This is principally determined by the relative permittivity of the concrete, which in turn is determined predominantly by the moisture content.

Because of the difficulties in interpretation, surveys are normally conducted by specialists who rely on practical experience and have a knowledge of the limitations of the technique in practical situations. For example, features such as voids can be particularly difficult to detect if located very deep or beneath a layer of closely spaced reinforcing steel. Neural networks or artificial intelligence have been used to help with the interpretation of complex radar traces.

Radar reflects most strongly off metallic objects or from the interface between two materials with widely differing permittivities. An air-filled void in dry concrete, which does not differ very strongly in permittivity from the concrete itself, can therefore be difficult to detect, especially if the void is small. The same void filled with water, however, would be much more easily detected. Water strongly attenuates a radar signal and, using a 1 GHz antenna, typical practical penetrations of around 500 mm have been achieved for dry concrete and 300 mm for water-saturated concrete. If the water is contaminated with salt, penetration is likely to be smaller still.

Acoustic emission

Theory
As a material is loaded, localized points may be strained beyond their elastic limit, and crushing or micro-cracking may occur. The kinetic energy released will propagate small-amplitude elastic stress waves throughout the specimen. These are known as acoustic emissions, although they are generally not in the audible range, and may be detected as small displacements by transducers positioned on the surface of the material.

An important feature of many materials is the Kaiser effect, which is the irreversible characteristic of acoustic emission resulting from applied stress. This means that if a material has been stressed to some level, no emission will be detected on subsequent loading until the previously applied stress level has been exceeded. This feature has allowed the method to be applied most usefully to materials testing, but unfortunately the phenomenon does not always apply to plain concrete. Concrete may recover many aspects of its pre-cracking internal structure within a matter of hours owing to continued hydration, and energy will again be released during reloading over a similar stress range. More recent tests on reinforced concrete beams have shown that the Kaiser effect is observed when unloading periods of up to 2 h have been investigated. However, it is probable that over longer time intervals the autogenic 'healing' of micro-cracks in concrete will negate the effect.

Equipment
Specialist equipment for this purpose is available in the UK as an integrated system in modular form and lightweight portable models may be used in the field. The results are most conveniently considered as a plot of emission count rate against applied load (Figure 4.5).

Applications and limitations
As the load level on a concrete specimen increases, the emission rate and signal level both increase slowly and consistently until failure approaches, and there is then a rapid increase up to failure. While this allows crack initiation and propagation to be monitored during a period of increasing stress, the method cannot be used for either individual or comparative measurement under static load conditions. Mature concrete provides more acoustic emission on cracking than young concrete, but this confirms that emissions do not show a significant increase until about 80–90% of ultimate stress. The absence of the Kaiser effect for concrete effectively rules out the method for establishing a history of past stress levels. Hawkins et al.,[13] however, described laboratory tests which indicate that it may be possible to detect the degree of bond damage caused by prior loading, if the emissions generated by a reinforced specimen under increasing

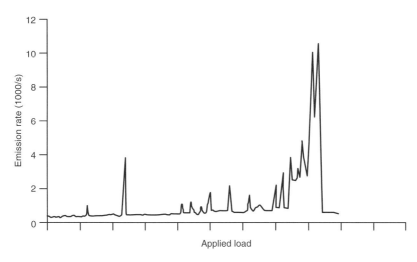

Figure 4.5 *Acoustic emission plot.*

load are filtered to isolate those caused by bond breakdown, since debonding of reinforcement is an irreversible process. Titus *et al.*[14] also suggested that it may be possible to detect the progress of micro-cracking due to corrosion activity. Long-term creep tests at constant loading by Rossi *et al.*[15] showed a clear link between creep deflection, essentially caused by drying shrinkage micro-cracking, and acoustic emission levels. However, in tests on plain concrete beams, together with fibre-reinforced and conventional steel-reinforced beams, Jenkins and Steputat[16] concluded that acoustic emission gave no early warning of incipient failure.

The application to concrete of acoustic emission methods has not yet been fully developed, and as equipment costs are high they must be regarded as essentially laboratory methods. However, there is clearly future potential for use of the method in conjunction with *in situ* load testing as a means of monitoring cracking origin and development and bond breakdown, and to provide a warning of impending failure.

Infrared thermography

This technique uses infrared photographs taken from a structure which has been heated, as it cools. The heating is normally performed by the sun in daytime and the photographs are best recorded in the evening as the structure cools. Infrared thermography offers many potential advantages over other physical methods for the detection of delamination in bridge decks. Areas of sound and unsound concrete will exhibit different thermal characteristics and thus have different surface temperatures as the structure cools. Delaminated areas, for example, will have a different temperature gradient to sound areas. Water-saturated concrete will appear quite different to dry concrete. The temperature differences are small, however, and cannot readily be recorded on infrared film. To view the small temperature differences, a cathode ray tube display is used with the different temperatures recorded as different shades of grey. Thermal contours can, however, be automatically superimposed and colour monitors can be used to give a more graphic picture.

The surface of the structure needs to be viewed from a reasonable distance and so cannot be recorded while, for example, standing on a deck. Some success has been achieved working from a vehicle at a height of 6 m, provided that the temperature differences were at least 2°C. Working from an aircraft or helicopter avoids the need for lane closures, but has not consistently shown good results.

Holt and Eales[17] described the successful use of thermography to evaluate effects in highway pavements with an infrared scanner and coupled real-time video scanner mounted on a 5 m high mast attached to a van. This is driven at up to 25 km/h and images are matched by computer. Procedures for infrared thermography in the investigation of bridge deck delamination are given in ASTM D4788.

Hidden voids or ducts can also sometimes be detected, and techniques have been developed to detect reinforcing bars that have been heated by electrical induction. The more recent development of 12 bit equipment[18] has improved the sensitivity to within ±0.1°C. This has enabled high-definition imaging and accurate temperature measurement on buildings. The smallest detectable area is reported to be 200×200 mm.

Infrared thermography can also be used to reduce heat losses at hot spots by identifying missing thermal insulation.

Testing for reinforcement corrosion

Half-cell potential testing

Steel embedded in good quality concrete is protected by the high alkalinity pore water which, in the presence of oxygen, passivates the steel. The loss of alkalinity due to carbonation of the concrete or the penetration of chloride ions (arising from either marine or de-icing salts, or in some cases present *in situ* from the use of a calcium chloride additive) can destroy the passive film.[19–22] In the presence of oxygen and humidity in the concrete, corrosion of the steel starts. A characteristic feature for the corrosion of steel in concrete is the development of macro-cells, that is the co-existence of passive and corroding areas on the same reinforcement bar forming a short-circuited galvanic cell, with the corroding area as the anode and the passive surface as the cathode. The voltage of such a cell can reach as high as 0.5 V or more, especially where chloride ions are present. The resulting current flow (which is directly proportional to the mass lost by the steel) is determined by the electrical resistance of the concrete and the anodic and cathodic reaction resistance.[23]

The current flow in the concrete is accompanied by an electrical field which can be measured at the concrete surface, resulting in equipotential lines that allow the location of the most corroding zones at the most negative values. This is the basis of potential mapping, the principal electrochemical technique applied to the routine inspection of reinforced concrete structures.[24,25] The use of the technique is described in an American Standard, ASTM C876-80, Standard Test Method for Half Cell Potentials of Reinforcing Steel in Concrete.

Factors affecting the potential field

When surface potentials are taken, they are measured remote from the reinforcement because of the concrete cover. The potentials measured are therefore affected by the ohmic drop, i.e. the potential drop in the concrete. Several factors have a significant effect on the potentials measured.

- **Concrete cover depth:** with increasing concrete cover, the potential values at the concrete surface over actively corroding and passive steel become similar. Thus, locating small corroding areas becomes increasingly difficult.
- **Concrete resistivity:** the concrete humidity and the presence of ions in the pore solution affect the electrical resistivity of the concrete. The resistivity may change both across the structure and with time as the local moisture and salt content vary. This may create an error of $\pm 50\,mV$ in the measured potentials.[26]
- **Highly resistive surface layers:** the macrocell currents tend to avoid highly resistive concrete. The measured potentials at the surface become more positive and corroding areas may be undetected.[25]
- **Polarization effects:** steel in concrete structures immersed in water or in the earth often have a very negative potential as a result of restricted oxygen access.[27] In the transition region of the structure (splash zone or above ground), negative potentials can be measured owing to galvanic coupling with immersed rebars. These negative potentials are not related to corrosion of the reinforcement.

Procedure for measurement

To measure half-cell potentials, an electrical connection is made to the steel reinforcement in part of the member to be assessed (Figure 4.6). This is connected to a high-impedance digital millivoltmeter, often backed up with a data-logging device. The other connection to the millivoltmeter is taken to a copper/copper sulfate ($Cu/CuSO_4$) half cell, which has a porous connection at one end which can be touched to the concrete surface. This

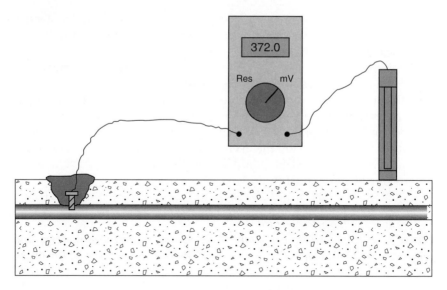

Figure 4.6 *Half-cell potential measurement. The voltage between a half cell (right) and a direct connection to the rebar (left) is measured with a high-resistance voltmeter.*

will then register the corrosion potential of the steel reinforcement nearest to the point of contact. Silver/silver chloride half cells are also used which are more robust; they give a potential approximately 50 mV more positive than a copper/copper sulfate half cell. By measuring results on a regular grid and plotting results as an equipotential contour map, areas of corroding steel may readily be seen. Using three-dimensional mapping techniques, a more graphical representation of the corrosion can be shown.

Results and interpretation
According to the ASTM method, corrosion can only be identified with 95% certainty at potentials more negative than -350 mV Cu/CuSO$_4$. Experience has shown, however, that passive structures tend to show values more positive than -200 mV and often positive potentials. Potentials more negative than -200 mV may be an indicator of the onset of corrosion. The patterns formed by the contours can often be a better guide in these cases.

In any case, the technique should never be used in isolation, but should be coupled with measurement of the chloride content of the concrete and its variation with depth, and also the cover to the steel and the depth of carbonation.

Resistivity

The electrical resistivity is an indication of the amount of moisture in the pores, and the size and tortuosity of the pore system. Resistivity is strongly

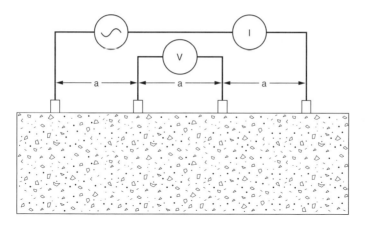

Figure 4.7 *Wenner probe. The four probes are equally spaced at a distance* a. *An imposed ac signal is applied and the voltage difference between the central probes noted.*

affected by concrete quality, i.e. cement content, w/c ratio, curing and additives used.

Equipment and use
The main device in use is the four-probe resistivity meter. This instrument has been modified from soil applications and is used by pushing probes directly onto the concrete with moisture or gels to enhance the electrical contact. Millard *et al.*[28] described two versions of the equipment. In the case of the four-probe Wenner array, the resistivity ρ of the concrete is given by the equation:

$$\rho = 2\pi\, a\frac{V}{I} \quad \text{(Figure 4.7)}$$

Some variations use drilled-in probes or a simpler, less accurate two-probe system. An alternative approach measures the resistivity of the cover concrete by a two-electrode method using the reinforcing network as one electrode and a surface probe as the other.[29] Concrete resistivity of the area around the sensor is obtained by the formula:

$$\rho = 2\pi RD \; (\Omega \text{ cm})$$

where R is the resistance by the '*IR* drop' from a pulse between a surface electrode and the rebar network measured by a half-cell reference electrode, and D is the electrode diameter of the sensor.

Interpretation is empirical. Examples of typical values are given in Appendix 1.

Limitations

The resistivity measurement is a useful additional measurement to aid in identifying problem areas or confirming concerns about poor-quality concrete. Readings can only be considered alongside other measurements.

There is a frequent temptation to multiply the resistivity by the half-cell potential and present this as the corrosion rate. This is incorrect. The corrosion rate is usually controlled by the interfacial resistance between the steel and the concrete, not the bulk concrete resistivity. The potential measured by a half cell is not the potential at the steel surface that drives the corrosion cell. Any correlation is fortuitous, as described above. However, a high-resistivity concrete will not sustain a high corrosion rate whereas a low-resistivity concrete can, if the steel is depassivated and there is sufficient oxygen and moisture present, and if the steel has been depassivated by the presence of chlorides or carbonation.

Corrosion rate

The corrosion rate is probably the nearest the engineer can come to measuring the rate of deterioration with current technology. There are various ways of measuring the rate of corrosion, including ac impedance and electrochemical noise.[30] However, these techniques are not field-worthy for the corrosion of steel in concrete, so this section will concentrate on linear polarization, also known as polarization resistance, as used in the GECOR device, available from the Spanish firm Geocisa.

Property to be measured

It is possible, with varying degrees of accuracy, to measure the amount of steel dissolving and forming oxide (rust). This is done directly as a measurement of the electric current generated by the anodic reaction:

$$Fe \rightarrow Fe^{2+} + 2e^-$$

and consumed by the cathodic reaction:

$$H_2O + \tfrac{1}{2}O_2 + 2e^- \rightarrow 2OH^-$$

and then converting the current flow by Faraday's law to metal loss:

$$m = \frac{MIt}{zF}$$

where m = mass of steel consumed, I = current (amperes, A), t = time (seconds, s), F = 96 500 A·s, z = ionic charge (2 for Fe $\rightarrow$ Fe^{2+} + 2e$^-$), and M = atomic mass of metal (56 g for Fe). This gives a conversion of $1\,A\cdot cm^{-2} = 11.6\,m$ per year.

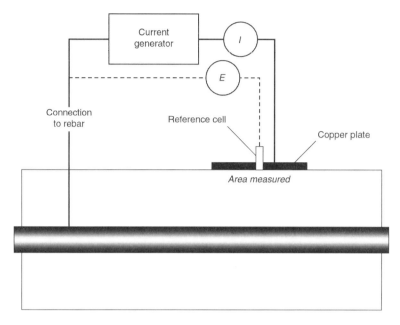

Figure 4.8 *Schematic of corrosion rate measuring device.*

Equipment and use: linear polarization

The system has a connection to the reinforcement, a half cell, an auxiliary electrode to apply the perturbing current and a battery-operated unit to supply the dc electric field and measure its effect via the half cell. The simplest device of this type is shown in Figure 4.8.

The potential is measured from the central half cell. Current I is passed from the surrounding electrode to the steel and the potential shift E is measured. This can be repeated for increasing increments of I. The potential must be stable throughout the reading so that a true E is recorded.

The equation for the corrosion current is given by Stern and Geary:[31]

$$I_{corr} = \frac{B}{R_P}$$

where I_{corr} is the corrosion current and B is a constant related to the anodic and cathodic Tafel slopes and the polarization resistance $R_P = \delta E/\delta I$, I being the change in current and E being the change in potential.

The problem with the simple devices is that they can be slow to operate and, more importantly, they do not define the area of measurement accurately. At low corrosion rates this can lead to errors by orders of magnitude.[32]

The GECOR device is more accurate. This works on the linear polarization principle and uses a guard ring. The device has been described in several papers[33] and the developers have worked on the interpretation of results from carbonated as well as chloride-induced corrosion. Its important features are the two extra half cells that are used to control the guard ring current and define the area of measurement. The most recent version of the GECOR device, the GECOR8 has the ability to perform potential mapping and can determine whether a bar is passive even when in a cathodic protection system with the current switched on.

In an assessment of three different devices, one without a guard ring, one with a simple guard ring and one with a sophisticated half-cell controlled guard ring, the researchers found good correlation between these devices and the most sophisticated laboratory measurements, except where the concrete resistivity was very high or cover to the rebar was very deep. Independent field trials also showed good performance.[32]

Gowers *et al.*[34] used the linear polarization technique with embedded probes (half cell and a simple counter-electrode) to monitor the corrosion of marine concrete structures. This technique was described previously without reference to isolating the section of bar to be measured.[35] By repeating the measurement in the same location on an isolated section of steel of known surface area, the corrosion rate of the actual rebar can be inferred. The main problem is the long-term durability of electrical connections in marine conditions.

Interpretation: linear polarization

The following broad criteria for corrosion have been developed from field and laboratory investigations with the sensor-controlled guard ring device:

$$I_{corr} < 0.1 \, \mu A/cm^2 \quad \text{passive condition}$$
$$I_{corr} \; 0.1–0.5 \, \mu A/cm^2 \quad \text{low to moderate corrosion}$$
$$I_{corr} \; 0.5–1 \, \mu A/cm^2 \quad \text{moderate to high corrosion}$$
$$I_{corr} > 1 \, \mu A/cm^2 \quad \text{high corrosion rate}$$

These measurements are affected by temperature and relative humidity (RH), so the conditions of measurement will affect the interpretation of the limits defined above. The measurements should be considered accurate to within a factor of two.

Work has been done in translating I_{corr} to section loss and end of service life.[36] However, the loss of concrete is the most usual cause for concern, rather than loss of reinforcement strength. It is far more difficult to predict cracking and spalling rates, especially from an instantaneous measurement. A simple extrapolation assuming that the instantaneous corrosion rate on a certain day is the average rate throughout the life of the structure often gives inaccurate results. This is true for section loss. Converting that to delamination rates is even less accurate as it requires further assumptions about oxide volume and stresses required for cracking the concrete.

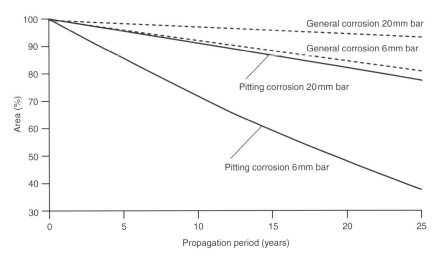

Figure 4.9 *Loss of cross-sectional area of reinforcement for* $I_{corr} = 1\,\mu A/cm^2$. *The high rate of pitting corrosion is clearly seen from the graph.*

The conversion of I_{corr} measurements to damage rates is still being worked on. Another area of concern is when corrosion is due to pitting. Much of the research on linear polarization has been done on highway bridge decks in the USA, where chloride levels are high and pitting is not observed. However, in Europe, where corrosion is localized in run-down areas on bridge substructures and pitting is more common, the problems of interpretation are complicated as the I_{corr} reading is coming from isolated pits rather than uniformly from the area of measurement. Work has been carried out to assess the damage that would be caused by both generalized and pitting corrosion, assuming a worst-case scenario for pitting.[37]

Laboratory tests with the guard ring device have shown that the corrosion rate can be up to 10 times higher than generalized corrosion.[38] This means that the device is very sensitive to pits. However, it cannot differentiate between pitting and generalized corrosion. Figure 4.9 illustrates the relationship between loss of cross-sectional area due to corrosion with respect to time for 6 mm and 20 mm bars. It can be seen that pitting is a far more serious type of corrosion than general corrosion.

On-site chemical testing

Two chemical tests can be carried out *in situ* to determine chemical modifications to concrete, both of which have immediate significance in determining the risk of corrosion to the reinforcement.

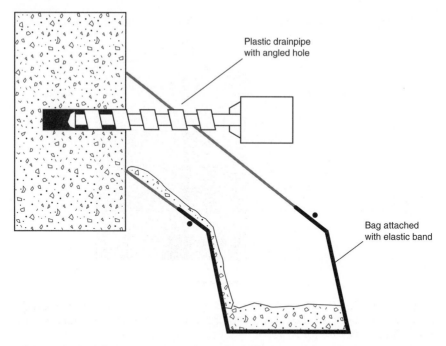

Plastic drainpipe
with angled hole

Bag attached
with elastic band

Figure 4.10 *Method for taking samples for chloride analysis.*

Chloride content

A concrete face exposed to concrete will gradually be permeated with chloride ions. In an investigation it is of interest to know the depth to which these ions have penetrated, as this measurement permits the estimation of the time it will take before the rebar depassivates and corrosion starts. The simplest way in which to measure chloride content is to drill a hole into the concrete, collect the dust and then test the recovered material for chloride (Figure 4.10). Using a technique such as this, it is possible to obtain a chloride profile through the cover zone of the concrete by drilling and testing in a series of steps. The chloride level in the concrete can be measured either in the laboratory or by field test methods. There are several rapid field test methods now available such as Quantab and Hach chloride test kits (chemical analytical methods are discussed in Chapter 5). On-site tests using the simple test kits should be taken as indicative only and are no substitute for correctly performed laboratory tests.

Carbonation

When concrete carbonates it undergoes a reduction of pH from about 13 to 8. To carry out a site determination of the depth to which concrete is

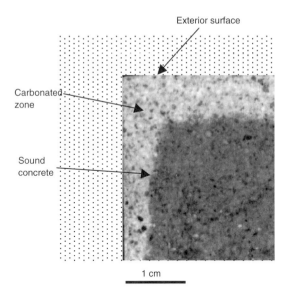

Figure 4.11 *Test for carbonation. A freshly fractured sample of mortar is sprayed with phenolphthalein solution. The darker area at the centre of the specimen is stained red, indicating that the pH is over 8.*

carbonated, it is necessary to break out a sample of the concrete: the drilling technique referred to above will tend to overestimate the alkalinity of the concrete owing to the possibility of exposing unhydrated cement. The test for carbonation is based on the measurement of pH with an indicator, a solution of phenolphthalein in alcohol. A freshly broken sample of the concrete is sprayed with the solution, the sound concrete is stained to a pink colour and the depth to which the pH of the concrete has been modified can be simply measured (Figure 4.11). The speed of the colour is also important; a slow colour can indicate partial carbonation. (See also chemical testing in Chapter 5.)

Strictly speaking, this test is a measure of the depth to which the alkalinity of the concrete has been reduced. Carbonation is the most common reason for loss of alkalinity; however, this phenomenon may result from other attacks by chemicals such as ammonium salts.

On-site permeability measurement

The type of permeability test used should match the problem being addressed. If the problem is carbonation, for example, which is driven by the diffusion of carbon dioxide gas, it is logical to use a gas permeability test. For field testing there are only two tests that can be easily applied, the initial surface absorption test (ISAT) and the Figg test. Both of these tests can be used in the laboratory and on site and can be used for quality control and durability monitoring.

The ISAT[39] measures the unsteady, unsaturated flow of water through concrete. A sealed chamber is clamped to the surface of the concrete and charged with water at a low pressure of 200 mm. After stabilization, the surface absorption is determined by the time required for a fixed volume of water to be absorbed.

The Figg test[40,41] can be used to measure air and water permeability. A 5.5 mm diameter hole is drilled into the concrete to a depth of 30 mm. The hole is plugged and water introduced through a syringe, and the permeability is measured using the time taken to absorb a fixed volume of water. The apparatus can be adapted to measure gas permeability, which is achieved by reducing the pressure in the drilled hole and timing the pressure recovery between two limits.

One problem with both of these tests is that the results will be dependent on the initial moisture content of the concrete.

Taking samples

Having made a superficial assessment of the condition of a structure it may be necessary to make a more detailed assessment of the condition of the concrete. Samples are normally taken by cutting cores with a diamond-edged coring cylinder. Cores are generally 100 mm in diameter and about 300 mm long; a sample of this size can be used for both physicochemical analysis and destructive strength testing. Larger size cores may have to be taken when the aggregate size is 30 mm or greater. The number of cores to be taken should be considered carefully as the cost of coring is relatively high. However, a statistically significant sample is required, a sample of one being potentially misleading. Guidance for the number of samples can be found in BS 1881: Part 124: 1988. A recovered sample will start to undergo modifications as soon as it removed; for this reason, samples should be sealed soon after removal. Once the core is removed from the concrete the core hole should be inspected to examine voids and breaks seen in the core.

Bibliography: ultrasound

British Standards Institution. *BS 4408: Recommendations for non-destructive methods of test for concrete*. Part 5. Measurement of the velocity of ultrasonic pulses in concrete. BSI, 1974.

Cement and Concrete Association. The ultrasonic-pulse-velocity method of test for concrete in structures. *Advisory Data Sheet No. 34*. Cement and Concrete Association, October 1977.

Chefdeville J. and Dawance G. L'auscultation dynamique du béton. *Annales de l'Institut Technique du Bâtiment et des Travaux Publics*, No. 140, July–Aug. 1950.

Chung H.W. An appraisal of the ultrasonic pulse technique for detecting voids in concrete. *Concrete* 1978, **12** (11).

Davis S.G. and Martin S.J. *The quality of concrete and its variation in structures.* Cement and Concrete Association, Technical Report 42.487, Nov. 1973.

Davis W.R. and Brough R. Ultrasonic techniques in ceramic research and testing. *Ultrasonics*, May 1972.

Drysdale R.G. Variations of concrete strength in existing buildings. *Magazine of Concrete Research*, 1973, **25** (85, Dec.).

Elvery R.H. Non-destructive testing of concrete and its relationship to specifications. *Concrete (Journal of the Concrete Society)* 1971, **5** (4).

Elvery R.H. Estimating strength of concrete in structures. Current Practice Sheet No. 10. *Concrete (Journal of the Concrete Society)* 1973, **7** (11).

Elvery R.H. and Din N. Ultrasonic inspection of reinforced concrete flexural members. *Symposium on non-destructive testing of concrete and timber*, Institution of Civil Engineers, London, June 1969, pp. 51–58.

Elvery R.H. and Forrester J.A. Non-destructive testing of concrete. *Progress in construction science and technology*. Medical and Technical Publishing, Aylesbury, 1971.

Elvery R.H. and Nwokoye D.N. Strength assessment of timber for glued laminated beams. *Symposium on non-destructive testing of concrete and timber*, Institute of Civil Engineers, London, June 1969, pp. 105–113.

Jones R. A review of the non-destructive testing of concrete. *Symposium on non-destructive testing of concrete and timber*. Institution of Civil Engineers, London, June 1969, pp. 1–7.

Jones R. and Gatfield E.N. Testing concrete by an ultrasonic pulse technique. Road Research Laboratory, *Technical Paper No. 34*. HMSO, London, 1955.

Lee I.D.G. Testing for safety in timber structures. *Symposium on non-destructive testing of concrete and timber*. Institution of Civil Engineers, London, June 1969, pp. 115–118.

Lee I.D.G. *Non-destructive testing of timber helicopter rotor blades*. Timber Research and Development Association. Test Record E/TR/15.

Leslie J.R. and Cheesman W.J. An ultrasonic method of studying deterioration and cracking in concrete structures. *Proceedings of the American Concrete Institute* 1949, **46**, 17–36.

Miles C.A. and Cutting C.L. Changes in the velocity of ultrasound in meat during freezing. *Journal of Food Technology* 1974, **9**, 119–122.

Samarrai M.A. and Elvery R.H. The influence of fibres upon crack development in reinforced concrete subject to uniaxial tension. *Magazine of Concrete Research* 1974, **26** (89, December).

Tomsett H.N. *The* in situ *evaluation of concrete using pulse velocity differences*. Cement and Concrete Association, July 1979.

Watkeys D.G. Non-destructive testing of concrete subject to fire attack. M.Sc. thesis, University College, London, 1967.

Whitehurst E.A. Soniscope tests concrete structures. *Proceedings of the American Concrete Institute*, 1951, **47**, 433.

References

1 Concrete Society. Diagnosis of deterioration in concrete structures: identification of defects, evaluation and development of remedial action. *Technical Report 54*. The Concrete Society.

2 American Concrete Institute. Guide for making condition surveys of concrete in service. *ACI 201.2R Manual of concrete practice Part 1*. American Concrete Institute, New York, 1991.

3 American Concrete Institute. *Guide for concrete inspection. ACI 311.R-95*. American Concrete Institute, New York, 1999.

4 Pullar-Strecker P. *Corrosion damaged concrete – assessment and repair*. Butterworth, London, 1987.

5 Rendell F. Underwater inspection. In *Underwater concreting and inspection*, ed. A. McLeish. Edward Arnold, London, 1994.

6 Rilem TC104. Draft recommendation for damage classification of concrete structures. *Materials and Structures* 1994, **27** (170), 362–369.

7 BS 1881:204:1988. *Recommendations on the use of electromagnetic cover measuring devices*. British Standards Institution, 1988.

8 Aldred J.A. at Protovale (Oxford) Ltd. Quantifying the losses in covermeter accuracy due to congestion of reinforcement. *Proc. 5th Int. Conf. on Structural Faults and Repair*, 1993. Edinburgh Engineering Technics Press, Edinburgh.

9 Jones R. The non-destructive testing of concrete. *Magazine of Concrete Research* 1949, (No. 2, June) 67–78.

10 Jones R. *Non-destructive testing of concrete*. Cambridge University Press, Cambridge 1962.

11 Jones R. and Gatfield E.N. *Testing concrete by an ultrasonic pulse technique*. Road Research Laboratory, Technical Paper No. 34. HMSO, London, 1955.

12 Facaoaru I. Non-destructive testing of concrete in Romania. *Symposium on non-destructive testing of concrete and timber*, Institution of Civil Engineers, London, June 1969, pp. 39–49.

13 Hawkins N.M., Kobayashi A.S. and Forney M.E. Use of holographic and acoustic emission techniques to detect structural damage in concrete members. *Experimental methods in concrete structures for practitioners*. ACI, Detroit, MI, 1979.

14 Titus R.N.K., Reddy D.W., Dunn S.E. and Hartt, W.H. Acoustic emission crack monitoring and prediction of remaining life of corroding reinforced concrete beams. *Proc. 4th European Conference on NDT*, Vol. 2. Pergamon Press, Oxford, 1988, pp. 1031–1040.

15 Rossi P., Goddart N., Robert J.L., Gervais J.P. and Bruhart D. Investigations of the basic creep of concrete by acoustic emission. *Materials and Structures* 1994, **27**, 510–514.

16 Jenkins D.R. and Steputat C.C. Acoustic emission monitoring of damage initiation and development in structurally reinforced concrete beams. *Proc. Structural Faults and Repair '93*, Vol. 3. Engineering Technics Press, Edinburgh 1993, pp. 79–87.

17 Holt F.B. and Eales J.W. Non-destructive evaluation of pavements. *Concrete International* 1987, **9** (6) 41–45.

18 Buyukozturk O. Imaging of concrete structures. *NDT&E International* 1999, **31**, 233–243.

19 American Concrete Institute, Committee 222. Corrosion of metals in concrete. ACIR-85. American Concrete Institute, Detroit, MI, 1985.

20 Page C.L. and Treadaway K.W.J. Aspect of the electrochemistry of steel in concrete. *Nature* 1982, **297**, 109.

21 Schiessl P. Corrosion of steel in concrete, RILEM Technical Committee 60-CSC, State of the Art Report.

22 Arup H. The mechanism of protection of steel by concrete. In *Corrosion of reinforcement in concrete construction*, ed. A.P. Crane. Society of Chemical Industry, London, 1983, pp. 151–157.

23 Elsener B. and Bohni H. *Schweiz, Ingenieur und Architekt*, 1987, **105**, 528.

24 Stratfull R.F. *Corrosion. NACE* 1957, **13**, 173t.

25 Berkeley K.G.C. and Pahmanaban S. Practical potential monitoring in concrete. *GCC Seminar on Corrosion*. Telford, London, 1987, pp. 115–131.

26 John D.G., Eden D.A., Dawson J.L. and Langford P.E. *Proc. Conf. Corrosion/87*, San Francisco, CA, 9–13 Mar. 1987.

27 Popovics S., Simeonov Y., Bozhinov G. and Barovsky N. In *Corrosion of reinforcement in concrete construction*, ed. A.P. Crane. Society of Chemical Industry, London, 1983, pp. 193–222.

28 Millard S.G., Harrison, J.A. and Gowers K.R. Practical measurement of concrete resistivity. *British Journal of NDT* 1991, **33** (2), 59–63.

29 Newman J. Resistance for flow of current to a disk. *Journal of the Electrochemical Society* 1966, **113**, 501–502.

30 Dawson J.L. Corrosion monitoring of steel in concrete. In *Corrosion of reinforcement in concrete construction*, ed. A.P. Crane. Ellis Horwood, for Society of Chemical Industry, London, 1983, pp. 175–192.

31 Stern M. and Geary A.L. Electrochemical polarisation. I. A theoretical analysis of the shape of polarisation curves. *Journal of the Electrochemical Society* 1957, **104**, 56–63.

32 Fliz J., Sehgal D.L., Kho Y.-T., Sabotl S., Pickering H., Osseo-Assare K. and Cady P.D. Condition evaluation of concrete bridges relative to reinforcement corrosion, Vol. 2. *Method for measuring the corrosion rate of reinforcing steel. SHRP-S-324*. National Research Council, Washington, DC, 1992.

33 Feliú S., González J.A., Andrade C. and Feliú V. On-site determination of the polarisation resistance in a reinforced concrete beam. *Corrosion* 1987, **44** (10), 761–765.

34 Millard S.G., Gowers K.R. and Gill J.S. Reinforcement corrosion assessment using linear polarization techniques. In *Evaluation and rehabilitation of concrete structures and innovations in design*, ed. V.M. Malhotra, Proc. of Int. Conf., Hong Kong, 1991, SP-128, Vol. 1. Detroit ACI, 1992.

35 Langford P. and Broomfield J. Monitoring the corrosion of reinforcing steel. *Construction Repair* 1987, **1** (2), 32–36.

36 Andrade C., Alonso M.C. and González J.A. An initial effort to use corrosion rate measurements for estimating rebar durability. In *Corrosion rates of steel in concrete, ASTM STP 1065*, eds N.S. Berke, V. Chakar and D. Whiting. ASTM, Philadelphia, PA, 1990, pp. 29–37.

37   Rodriguez J., Ortega, L.M., Casal J. and Diez J.M. Corrosion of reinforce-
     ment and service life of concrete structures. *Proc. 7th Int. Conf. Durability of
     Building Materials and Components*, Stockholm, 1996.
38   González J.A., Andrade C., Rodríguez P., Alonso C. and Feliú S. Effects of
     corrosion on the degradation of reinforced concrete structures. *Progress in
     understanding and prevention of corrosion*. Institute of Materials for Euro-
     pean Federation of Corrosion, Cambridge University Press, Cambridge,
     1993, pp. 629–633.
39   British Standard Institution. BS 1881. Testing concrete part 208:1996.
     *Recommendations for the determination of the initial surface absorption of
     concrete*. BSI, London, 1996.
40   Figg J.W. Method of measuring the air and water permeability of concrete.
     Magazine of Concrete Research 1973, **25** (85), 213–219.
41   Cather R., Figg J.W., Marsden A.F. and O'Brien T.P. Improvements to the
     Figg method for determining the air permeability of concrete. *Magazine of
     Concrete Research* 1984, **36** (129), 241–245.

Chapter 5

Laboratory testing

This chapter describes some of the laboratory testing used to diagnose problems with concrete. This phase of a study will follow the site inspection and *in situ* testing. At this point in the investigation much evidence will have been assembled and the laboratory study will be looked on to provide a more in-depth knowledge of the condition of the materials and whether there has been a modification in their properties. The results of these studies are vital in the diagnosis of problems and provide important evidence in the estimation of residual service life and the development of remedial work planning.

One of the key objectives of this book is to describe the role of X-ray diffraction (XRD) analysis and scanning electron microscopy (SEM) in the examination of concrete. Before looking at these methods it is important to place these more advanced techniques into context. The normal route of laboratory testing will commence with 'classical' methods, namely chemical testing and petrography. These more classical techniques enable an overview of the problem to be sought, which will logically lead to the more advanced, and expensive, techniques being used in a more effective way. Laboratory testing will be conducted in the following order.

- Initial laboratory tests will start with a visual examination of the material, which leads to a petrographic analysis.
- Chemical testing may be required to determine the levels of chlorides, sulfates, etc. Confirmation of depth of carbonation can also be carried out.
- Strength and permeability testing can be carried out to examine the physical properties of the material.
- Where specific problem areas are identified the more advanced techniques, such as XRD and SEM, are brought to bear on the problem.

As the more modern techniques such as SEM linked to micro-analysis are developed, they are finding an ever wider industrial application in the examination of concrete.

Petrographic examination

The following section provides an overview of the subject of concrete petrography. For a full and authoritative description of the subject, see references 1 and 2.

Preliminary examination

The initial examination of samples is carried out to record dimensions and main features. The features observed include:

- the presence and position of reinforcement
- the extent to which reinforcement is corroded
- the nature of the external surfaces of the concrete
- the features and distribution of macro-cracks and fine cracks
- the distribution, size range and type of the aggregate
- the type and condition of the cement paste
- any superficial evidence of deleterious processes affecting the concrete.

Fresh fractured surfaces

After the recovery of the cores, specimens have to be selected from detailed analysis. When cutting the core for the preparation of polished surfaces and thin section samples, pieces of the cores that are surplus to requirement can be used for analysis of freshly fractured surfaces. Sections of the sample are broken to produce fresh surfaces. These surfaces can be examined with a binocular microscope and will provide information concerning the contents of voids, and the nature of aggregate surfaces and crack surfaces.

Polished surfaces

A plate is cut, where possible, from each sample. This is typically about 20 mm thick and usually provides as large a section of the sample as is possible. The plate is polished to give a high-quality surface that can be examined with a high-quality binocular microscope or even with the petrological microscope if necessary. For examinations up to a magnification of $\times 100$ the surface can be finely ground with a $5\,\mu m$ grit; for greater magnifications it will be necessary to polish the surface with a sub-micrometre medium. If the surface of the sample is friable, the surface must be stabilized with a resin. Consideration should be given to the lubrication used in the cutting and grinding of concrete; many components are soluble and will be eliminated during sample preparation if water is used. The polished surface samples can be produced relatively cheaply and can be used to assess the following features of the concrete:

- size, shape and distribution of coarse and fine aggregate
- relative abundance of rock types in the coarse aggregate
- coherence, colour and porosity of the cement paste
- distribution, size, shape and content of voids
- composition of the concrete, in terms of the volume proportions of coarse aggregate, fine aggregate, paste and void
- distribution of fine cracks and micro-cracks. Often the surface is stained with a penetrative dye so that these cracks can be seen. Micro-crack frequency is measured along lines of traverse on the surface and expressed as cracks/cm.

The technique of fluorescence microscopy can also be applied to polished surfaces. The method consists of exciting fluorescence in the specimen with short-wave light (ultraviolet). To achieve the fluorescence it is often necessary to stain the surface with a fluorescent dye. An important application of this technique is in the identification of alkali silica gel by staining with uranyl acetate acetic acid solution.[3,4]

Crack detection in samples always presents problems because of the possible crack formation in the cutting and drying of the sample. One possible method of avoiding the cracking induced in drying the sample is to impregnate the surface with dye dissolved in alcohol.[5] A range of staining techniques can be used for the identification of chemical species including sulfates, sulfides and carbonates (see reference 2, pp. 79–82).

Thin sections

The preparation of thin sections is expensive and should be carried out by a skilled technician. Because of this, the selection of areas to be examined is carried out with some care and based on the observations made previously.

The section is usually made from a plate cut at right angles to the external surface of the concrete, so that the outer 70 mm or so of the concrete is included in the section. Sometimes it is more appropriate, where specific problems are being investigated, to make the section from inner parts of the concrete. The thin section specimens normally measure about 50 × 70 mm.

In manufacturing the thin section a plate some 10 mm thick is cut from the sample. This is impregnated with a penetrative resin containing a yellow fluorescent dye. The resin penetrates into cracks, micro-cracks and capillary pores in the sample. One side of the impregnated plate is then polished and the plate is mounted on to a glass slide. The surplus sample is then removed and the plate is ground and polished to give a final thickness between 20 and 30 μm. At all stages the cutting and grinding is carried out using an oil-based coolant to prevent further hydration of the cement and excessive heating of the section. The thin section is covered and then examined with a high-quality petrological microscope. The analysis of the thin sections can be used for the examination of the following features of the concrete.

- The rock types present in the coarse and fine aggregate and, in particular, details of structures within the aggregates can be identified.
- Details of the aggregate properties are measured, such as the degree of strain in quartz.
- The size, distribution and abundance of phases in the cement paste are assessed, including the occurrence of calcium hydroxide and the amount of residual unhydrated clinker.
- The presence of cement replacement phases such as slag or pulverized fuel ash (PFA) can usually be observed (however, the amount of these phases cannot be judged accurately by visual examination, using SEM methods estimation of slag and PFA content is possible). The presence of high alumina cement can be detected and the type of cement clinker can often be assessed.
- Many products resulting from the deterioration of either the cement paste or the aggregate can be recognized.

As with the examination of polished surfaces a range of selective staining techniques can be applied. Fluorescence microscopy, for example, can be used to identify areas of high porosity by the use of dye-impregnated resins. Fluorescence microscopy of thin sections can also be used to estimate the water/cement (w/c) ratio of a concrete.[2]

Composition of concrete

The composition of the sample is measured using either the polished slice or the thin section, depending on the size of the sample and on details of the aggregate type and paste. The thin section is preferable, for example, where large quantities of dust are present. The volume proportions are found by the method of point counting using a mechanical stage. The amount of coarse aggregate can also be assessed by this method if a distinction can be made between coarse and fine aggregate. The results obtained usually represent the sample reasonably, but may not represent the concrete.

The amount of individual rock types present in the aggregate as a whole is assessed and the saturated density of the sample is measured by the method of immersion in water using vacuum impregnation to ensure saturation. From this information and the volume proportions, the weight fractions of aggregate, cement and water can be calculated.

Water/cement ratio

The hydration processes of cement paste vary significantly with the original w/c ratio. Concretes with a low w/c ratio tend to leave substantial quantities of unhydrated cement clinker and to develop only limited amounts of coarsely crystalline calcium hydroxide. In particular, the extent to which calcium hydroxide is separated into layers on aggregate surfaces and occurs in voids and on void surfaces varies with the original w/c ratio. The number and proportion of unhydrated cement clinker particles varies inversely with the original w/c ratio. Comparison with standard concretes made with known w/c ratios, visually and by measurement, allows the w/c ratio of the cement paste to be assessed directly. The standard error attached to the estimation of w/c ratio by this means is considered to be approximately ± 0.03.

Further analysis

The application of advanced analytical techniques to concrete is becoming more widespread for the examination of surfaces and the identification of chemical phases. The techniques discussed in the later sections of this book discuss powerful tools that can complement and at times supersede the more classical methods developed in concrete technology. However, it must be appreciated that these advanced techniques are not a panacea for all problems of concrete analysis; the first analysis will inevitably involve a level of optical microscopy.

Concrete strength

Concrete strength has been the traditional characteristic used to specify, control and evaluate the condition of concrete. As previously discussed in this book, there is a strong correlation between strength, transport parameters and therefore durability. In the evaluation of the condition of an existing concrete structure some level of strength determination is essential. This information will be required in the event of a structural assessment to evaluate new performance levels for a structure. In Chapter 4 *in situ* strength assessments such as the rebound hammer and ultrasound were discussed. It can be difficult for these methods to yield a meaningful strength characteristic and therefore they require calibration against standard mixes. Once cores have been taken from a structure it is possible to repeat non-destructive strength tests in the laboratory; however, destructive testing is a more accepted, and transparent, method of testing.

Concrete will normally be specified as a 'design mix', meaning that the engineer will have asked for a concrete of a certain strength, either to ensure that he has adequate strength for structural purposes, or, more usually, to ensure the overriding requirement that the concrete has adequate durability. The concrete strength is quoted in terms of a cube strength, the compressive strength of a 100 or 150 mm cube of concrete (depending on the aggregate size). Other criteria are also used in the specification of concrete such as compressive strength of cylinders and tension tests. (Concrete is not a homogeneous material and therefore characteristics such as strength are accepted as having a wide band of variation.) Even with good quality control, a spread of results of $16\,N/mm^2$ is likely with a grade 40 concrete. This means that to ensure that most (95%) of the results are above $40\,N/mm^2$, the concrete supplier will aim for a target mean strength of some $48\,N/mm^2$.

When assessing the strength of existing concrete two problems are paramount; first, the statistical problem implicit in the characterization of an inhomogeneous material and, secondly, the problem of relating strength assessments on concrete cores to the original strength criteria used in the concrete specification.

The only solution to the statistical problem is to test sufficient samples to yield a statistically significant result. The second problem of strength criteria can be addressed within the core strength assessment protocol.

Concrete strength assessment

The results gained from concrete core samples will not be directly comparable with those from the original cube tests. Cube samples are fully compacted and have been stored under ideal curing conditions for 28 days prior to testing. Core samples, in contrast, have been taken from *in situ* concrete, cured in the structure, often inadequately, with perhaps less than perfect compaction. There are also settlement effects, with results on cores from the bottom of a wall or column differing by 15–30% from those of cores taken near the top.[6] Not surprisingly, there can be large differences in the strength measured on cores and on cubes made from the original concrete, as supplied.

Samples for assessment of compressive strength are cut from extracted cores. The length to diameter ratio of the cores should range from 1 to 2, with a preferred value being between 1 and 1.5. To ensure that the sample is correctly mounted in the test rig, the end of the sample should be trimmed or coated to provide two flat parallel end surfaces. This can be achieved by surfacing the samples with a mixture of sand, sulfur and carbon. Further information on the testing and interpretation of results can be found in the Concrete Society Technical Report 11[7] and Publication CS 126.[8] A supplement to Concrete Society Technical Report No. 11 to cover concrete made with ground-granulated blast furnace slag (GGBFS) and PFA has been published. An article in *Concrete* magazine summarizes the subject.[9]

The capped core is then crushed in a calibrated compression testing machine. The resulting failure load is converted first to a cylinder strength and secondly to an equivalent *in situ* cube strength. The cube strength f_{cube} corresponding to a cylinder compression f_{cyl} strength is given by the approximate relationship:

$$f_{cube} = 1.25 f_{cyl}$$

Determination of transport parameters

The two methods discussed in Chapter 4, the initial surface absorption test (ISAT) and the Figg test, can be used in the laboratory to confirm and strengthen field observations. The tests described in the following section are all laboratory-based tests that can be applied to sections of cores taken from a structure. The reader is reminded of the comment made earlier: the chosen test should represent the transport mechanism that is thought to be taking place in the concrete under investigation.

Water absorption or sorptivity

Water absorption and sorptivity tests have the virtue of being simple and very low cost.[10] The total water absorption test, BS 1881,[11] consists of drying a sample for 3 days at 105°C and then measuring the mass of water absorbed when it is immersed in water for 30 min. Typical absorption values are:

- high permeability concrete   >5%
- average concrete   3–5%
- low permeability concrete   <3%

The fundamental problems with this type of test are as follows.

- During drying at temperatures over 60°C there are changes to the cement paste structure and possibly micro-cracking.
- During the initial immersion there will be a non-representative uptake of water.
- The water absorption will be a function of the nature of the external surface of the concrete.

Sorptivity has been proposed as an alternative absorption method; in this case, a fluid moving into the concrete is controlled by capillary forces.[12] Dried samples are placed on a saturated sand bed and the mass increase in the concrete is recorded with respect to time for about 60 min. The slope of a graph of mass gain versus the square root of time can be used to characterize the capillary absorption of a concrete. This method has the benefit that the influence of the skin of concrete can be assessed even when suffering from heavy chemical attack.[13] Refinements of the method have proposed drying at 60°C until the mass of the sample becomes constant and then allowing the sample to cool to room temperature in a desiccator.[14,15]

Permeability

The water permeability of a concrete applies to the case where water permeates as a result of a pressure gradient. The low intrinsic water permeability demands that high-pressure differences are needed across the sample, typically $700\,kN/m^2$. Because of the low flow rates involved the test may take several days to stabilize and develop a steady flow. Once equilibrium has been established the permeability can be calculated using Darcy's law, i.e. from the flow rate and the pressure drop across the specimen. In a similar way, gas permeability can be measured. This test tends to be much quicker than water permeability but the results are dependent on the initial moisture content of the sample.

Diffusion

In the diffusion process a gas or an ionic species moves through a medium as a result of a concentration gradient. A typical experimental device consists of a disc of concrete separating two chambers, one containing distilled water and the other the solution in question. This type of test takes several weeks to reach equilibrium. Moreover, there is a potential problem due to surface blocking caused by the formation of chemical deposits, often formed when working with sea water.

Ionic diffusion rates can be accelerated by applying a voltage across the specimen (Figure 5.1). This can reduce the test time to hours rather than weeks. These techniques have been used successfully for determining diffusion characteristics of chlorides[16-19] and sulfates.[20,21]

Oxygen diffusion can be measured in a similar way to the chloride diffusion coefficients. Pure oxygen is passed over one face of the sample and nitrogen over the other, and the amount of oxygen transported through the

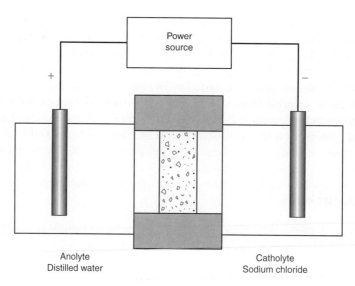

Figure 5.1   *Cell arrangement used for an accelerated chloride diffusion test.*

sample can be measured with a gas analyser. Oxygen diffusion correlates well with oxygen permeability, but it tends to be very sensitive to sample moisture content.

Chloride diffusion coefficients can be calculated from the observed chloride concentration profile found in the specimen. In principle, if the age of the structure is known and the profile concentration is measured it should be possible to determine the diffusion coefficient D by a curve-fitting technique from a solution of the diffusion equation:

$$C(x, t) = C_i + (C_s - C_i) erf\left(\frac{x}{(4tD)^{0.5}}\right)$$

where D is the diffusion coefficient, t is time, erf is an error function, and C_i is background concentration.

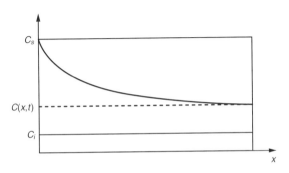

This approach has to be amended where chemical modification of the material occurs during the service life of the structure. In the case of marine structures there will be a significant alteration to the surface composition of concrete, leading to a diminution in the diffusion coefficient. This variation in diffusion coefficient is not linear with time. A semi-empirical approach to determine service life on the basis of chloride concentration distributions has been proposed by Helland *et al.*[22] It should also be noted that in real structures chloride penetration is also influenced by surface absorption.

Chemical tests

Chemical analysis of concrete can provide extremely useful information regarding the cause or causes of failure of concrete. The tests most frequently carried out include determination of chloride, sulfate and alkali content, determination of the depth of carbonation, and identification of the type of cement used in the original mix.

Chloride content

The generally accepted method of test for chloride in hardened concrete is described in BS 1881: Part 124.[23] The test involves crushing a sample of the concrete to a fine dust, extracting the chloride with hot dilute nitric acid and then adding silver nitrate solution to precipitate any chloride present. Ammonium thiocyanate solution is then titrated against the remaining silver and the amount of chloride determined from the difference between the added silver nitrate and that remaining after precipitating the chloride.

Faster and more precise methods based on ion-selective electrodes are now available.[24] Instrumental methods of analysis require accurate calibration with samples of known composition, preferably made from actual concrete or mortar. Test samples from these must be used regularly to confirm the correct functioning of the testing equipment.

Sulfate content

Sulfate is usually determined by the method given in BS 1881: Part 124: 1988. This involves an acid extraction and precipitation of the sulfate as barium sulfate with barium chloride solution. The resulting barium sulfate is filtered and weighed to determine sulfate gravimetrically.

Methods based on ion-selective electrodes and ion chromatography have also been used.

Depth of carbonation

The depth of carbonation can be measured on a freshly exposed section of the concrete, such as a core, by spraying with an indicator spray such as phenolphthalein. This turns pink when the concrete is alkaline (above pH 9.2), but remains colourless where the concrete is carbonated, usually as a more or less even zone extending to some depth from the surface. It should be noted that the pH at which the colour of phenolphthalein changes is lower than that at which passivity is lost (which occurs progressively below about pH 11). The test is described in a BRE information sheet.[25] It should be noted that carbonation along micro-cracks and along diffusion paths in poorly compacted concrete, or so-called reconstituted stone, may not be readily revealed by the phenolphthalein spray method. Petrographic methods can reveal carbonation of this kind and are recommended.

Cement content

The test to determine the cement content of concrete is given in BS 1881: Part 124: 1988. It requires the crushed concrete to be extracted with dilute acid and dilute alkali solution to remove the cement. The extract is then analysed for soluble silica and calcium oxide, which are the two major components (expressed as oxides) of Portland cement. The cement content is determined by simple proportion from the two parameters. Where soluble components from the aggregate interfere by contributing to the calcium content (e.g. if a limestone aggregate is present) then the silica value would be used for the determination of cement content. Conversely, if the silica value was inflated by some soluble component other than the cement, the lime value would be used, provided the analyst was confident that this was unaffected by soluble components from the aggregate. In practice, it is normal to analyse control samples of the aggregate, where these are available, to avoid these problems. With control samples, an accuracy of better than $\pm 25 \, kg/m^3$ is readily achievable.

Where cement replacement materials such as PFA and GGBFS are present, the situation is more complex. Nevertheless, accurate results can often be obtained using total analyses by, for example, X-ray fluorescence methods and applying simultaneous equations.[26]

High alumina cement (HAC)

At normal temperatures, the hydration of HAC results in the formation of hydrated calcium monoaluminate (CAH_{10}). Smaller amounts of C_2AH_8 and hydrous alumina are also formed. However, these hydrated calcium aluminates are meta-stable and can, at higher temperatures and in the presence of moisture, change to give the stable hydrated calcium aluminate C_3AH_6. This phenomenon is known as conversion and the amount of the change occurring as the degree of conversion. Following conversion, the increased porosity may permit rapid carbonation of the concrete, removing alkaline protection to the steel reinforcement, which may then suffer from corrosion.

A test was devised by the Building Research Station to show whether HAC is likely to be present in a concrete.[27] It essentially tests for a significant content of soluble aluminium in solution, following extraction with dilute sodium hydroxide solution. The presence of carbonate minerals renders any determination of the degree of conversion of the concrete potentially inaccurate. The best procedures for examination of HAC are petrographic and X-ray diffraction analyses.

References

1 French W.J. Concrete petrography: a review. *Quarterly Journal of Engineering Geology* 1991, **24**, 17–48.

2 St John D.A., Poole A.W. and Sims I. *Concrete petrography*. Arnold, London, 1998.

3 Natesaiyer K.C. and Hoover K.C. In situ identification of ASR products in concrete. *Cement and Concrete Research* 1988, **18**, 455–463.

4 Natesaiyer K.C. and Hoover K.C. Further study of an in situ identification for ASR products in concrete. *Cement and Concrete Research* 1989, **19**, 770–778.

5   Hornain H., Marchand J., Ammouche A., Commene J.P. and Moranville M. Microscopic observation of cracks in concrete – sample preparation technique using dye impregnation. *Cement and Concrete Research* 1996, **26** (4), 573–583.

6   British Standards Institution. BS 6089. *Guide to the assessment of concrete strength in existing structures.* BSI, 1981.

7   The Concrete Society. *Concrete core testing for strength.* Technical Report 11. The Concrete Society, Crowthorne, 1987, 50 pp.

8   The Concrete Society. *In situ concrete strength: an investigation into the relationship between core strength and standard cube strength.* CS 126. The Concrete Society, Crowthorne, 2000.

9   Grantham M.G. Concrete cube failures, counting the cost. *Concrete* 1993, Sept./Oct., 38–40.

10 Wilson M.A., Carter M.A. and Hoff W.D. British Standard and RILEM water absorption tests: a critical review. *Matériaux et Constructions* 1999, **32** (October), 571–578.

11 British Standards Institution. BS 1881: Part 122: 1983. *Method for determination of water absorption.* BSI, 1983.

12 McCarter W.J., Emerson M. and Ezirim H. Properties of concrete in the cover zone: developments in monitoring techniques. *Magazine of Concrete Research* 1995, **47** (172).

13 Camps J.-P., Jauberthie R. and Rendell F. The influence of surface absorption on sulphate attack. In *Concrete durability and repair technology.* International Conference, Creating with Concrete, Dundee, 6–10 Sept. 1999.

14 Sabir B.B., Wild S. and O'Farrell M. A water sorptivity test for mortar and concrete. *Matériaux et Constructions* 1998, **31** (212), 568–574.

15 Alexander M.G. Discussion of 'water sorptivity test for mortar and concrete'. *Matériaux et Constructions* 1999, **32** (Nov.), 695–696.

16 Whiting D. Rapid determination of the chloride permeability of concrete. Report no. FHWA-RD-81-119, 1991, NTIS BD No. 82140724.

17 Sergi G., Yu S.W. and Page C.L. Diffusion of chloride and hydroxile ions in cementitious materials exposed to a saline environment. *Magazine of Concrete Research* 1992, **44** (158), 63.

18 Andrade C. and Sanjuan M.A. Experimental procedure for the calculation of chloride diffusion coefficients in concrete from migration tests. *Advances in Cement Research* 1994, **6** (23), 127.

19 Andrade C., Castellote M., Cervigon D. and Alonso C. Influence of external concentration and testing time on chloride diffusion coefficient values. In *Corrosion of reinforcement in concrete construction*, eds C.L. Page, P.B. Bamforth and J.W. Figg. Proc. 4th Int. Symp. Royal Society of Chemistry, 1996, pp. 76–87.

20 Tumidajski P.J. and Tunc I. A rapid test for sulfate ingress into concrete. *Cement and Concrete Research* 1995, **25** (5), 924–929.

21 Tumidajski P.J., Chan G.W. and Philipose K.E. An effective diffusivity for sulphate transport into concrete. *Cement and Concrete Research* 1995, **25** (6), 1159–1163.

22 Helland S., Mangne M. and Carlsen J.E. Service life prediction of marine structures. In Concrete 95: towards better concrete structures *Concrete Institute of Australia,* Brisbane, Sept. 1995, Vol. 1, pp. 243–350.

23 British BS 1881: part 124. *Methods for analysis of hardened concrete.* BSI, 1988.

24 Grantham M.G. An automated method for the analysis of chloride in hardened concrete. *Proc. 1993 Conf. Structural Faults and Repair*, Edinburgh, June 1993. ECS Publications.

25 Building Research Establishment. Carbonation of concrete made with dense natural aggregates. *BRE Information Sheet IP 6/81.* HMSO, London, 1981.

26 Grantham M.G. Determination of slag and PFA in hardened concrete – the method of last resort revisited. In *Determination of the chemical and mineral admixture content of hardened concrete, ASTM STP 1253*, eds S.H. Kosmatka and A.A. Jeknavorian. American Society for Testing and Materials, Philadelphia, PA, 1994.

27 Building Research Establishment. A rapid chemical test for high alumina cement. *BRE Information Sheet IS15/74*, 1975.

Chapter 6

X-ray diffraction analysis

For background reading, see references 1–6 and Appendix 2.

Diffraction of X-rays

X-rays are electromagnetic radiation of very short wavelength (0.1–10 Å). If a crystal comprises a natural three-dimensional lattice, in which the spacing is of the same order as the wavelength of the X-ray, it will diffract within the lattice in the same manner as light or sea waves are diffracted. This is an elastic interaction between the X-ray and the solid, therefore the incident and diffracted rays have the same wavelength and energy. In 1913 the English physicist Sir W.H. Bragg and his son Sir W.L. Bragg studied why X-ray beams appeared to reflect from crystal surfaces at certain angles of incidence. This effect was attributed to X-ray wave interference, which is now known as X-ray diffraction (XRD). Bragg's law was proposed, which establishes the link between wavelength, incidence angle and the spacing between the planes of atoms in the atomic lattice.

Every species of crystal has a unique diffraction pattern that can enable identification and, consequently, can be used to detect the presence of a specific mineral in a mixture. Therefore, XRD analysis enables the identification of crystal phases and more specifically enables the determination of interplanar spacing between parallel planes, and thus the Miller indices of these planes; finally, it can be used for quantitative analysis.

Bragg's law

To gain an appreciation of the diffraction effect consider a collimated beam of monochromatic X-rays projected onto a sample at an incident angle of θ. The X-rays interacting elastically with an atom will be scattered in all directions, and most of these scattered rays will interfere destructively and eliminate each other. However, under certain conditions the diffracted rays will interfere constructively and give rise to a diffracted beam of X-rays.

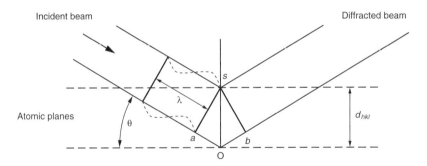

Figure 6.1   *Diffraction of an X-ray beam.*

A relationship exists between the wavelength of the X-ray λ, the incident angle of the beam θ and the distance between the planes of atoms (interplanar distance) d_{hkl} (Figure 6.1) (*hkl* being Miller indices, see Appendix 3).

If the diffracted rays are to be in phase, the distance *aob* must equal a multiple of the wavelength.

$$\text{Assume} \qquad aob = n\lambda$$
$$\text{Therefore} \qquad ao = n\lambda/2$$
$$\text{Angle} \qquad aso = bso = \theta$$

Consider the triangle *aso*:

$$\frac{ao}{os} = \sin\theta$$
$$n\lambda = 2d_{hkl}\sin\theta \tag{6.1}$$

where n is an integer. This is the basis for Bragg's law, where the angle θ is known as the Bragg angle.

It can be seen from this derivation that, for any wavelength and interplanar spacing, there will exist a number of diffracted rays where positive interference takes place. Figure 6.2 shows the variation in the intensity of the diffracted beam as the angle θ is varied from 0° to 90°.

X-ray diffraction analysis

X-ray generation

Figure 6.3 shows a typical layout of an X-ray tube that would be used in spectrometry. The cathode consists of a heated tungsten filament and the anode, the target, consists of a cooled metal block, often copper. Electrons are accelerated from cathode to anode by a potential difference capable of

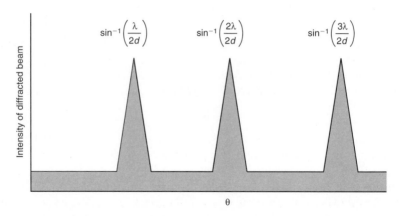

Figure 6.2 *Diffraction lines that would be observed in the case of a monochromatic X-ray, wavelength λ, being diffracted by two planes of atoms a distance d apart.*

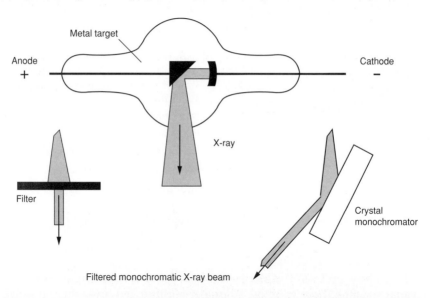

Figure 6.3 *Layout of an X-ray tube. The tube produces a polychromatic X-ray beam, the desired monochromatic beam is produced either by a filter (bottom left) or by using a crystal to diffract the beam (bottom right), thus enabling the desired wavelength to be selected at a pre-calculated diffraction angle.*

generating the desired characteristic peaks (e.g. $K\alpha$, $K\beta$, $L\alpha$) for the metal in the target, typically 20–60 kV. The X-rays produced are polychromatic and therefore a filter or monochromator is used to produce the monochromatic X-ray beam. This unit can be either an iron filter or a crystal inclined to the emitted source, such that the desired wavelength can be separated out by the XRD properties of the crystal (see Bragg's law).

The diffractometer

The diffraction spectrum for a material is obtained using a diffractometer, the basic layout of which is shown in Figures 6.4 and 6.5.

The monochromatic X-ray beam is directed at a material sample, where it is diffracted through the Bragg angle of θ:

$$2d_{hkl} \sin \theta = n\lambda$$

To ensure that the range of crystal forms and interplanar distances is scanned, the sample is ground to a powder, normally a sieve size less than $50\,\mu m$. The sample is inclined at an angle θ to the incident beam and a counter is positioned at an angle of 2θ to the incident beam (Figure 6.5). The sample is scanned by turning the sample at an angular velocity of ω and the counter at 2ω. The counter transforms the photons of X-ray radiation into an electrical pulse using a photoelectric effect; after amplification and filtering, these pulses are recorded to produce a trace of the intensity of the diffracted beam against 2θ.

A typical diffractogram used for mineral analysis is set out below:

- X-ray source: Cu $K\alpha$ specification of the X-ray source: copper target, and filtered to emit only the $K\alpha$ wavelength
- Step size: 0.005°   angular distance between each observation
- Count time: 0.05–1 s   time of observation.

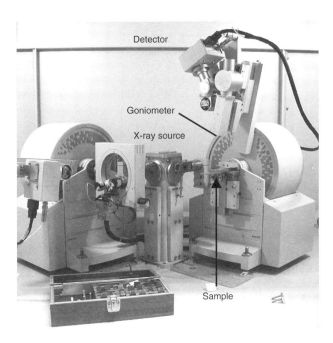

Figure 6.4 *Layout of a diffractometer.*

Interpretation of the XRD diagram

The diffraction diagram shown in Figure 6.6 shows the X-ray spectrum for calcium oxide. In general, the trace consists of two components:

- **a halo**, i.e. the wide band along the base of the trace, which is partly due to X-ray fluorescence from the amorphous material and partly due to electronic noise

Vertical goniometer

Figure 6.5   *Orientation of the X-ray source, the sample and the X-ray detector.*

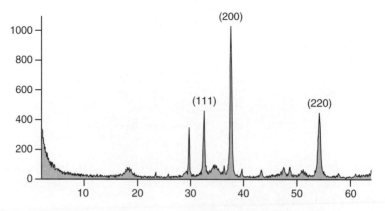

Figure 6.6   *Diffractogram for lime. This diagram shows the principal diffraction lines for lime, which have been referenced with the Miller indices corresponding to the diffraction planes. The other minor peaks present correspond to calcite, $CaCO_3$. Lime is normally obtained by the calcination (burning) of calcite; in the case of this sample the conversion is incomplete.*

- **peaks or diffraction lines**, which occur at angles and intensities that are unique to a mineral. The angle θ of a diffracted line can be linked to the interplanar distance between the family of planes in which the diffraction occurs. It is immediately evident from the trace that there are several peaks associated with the crystal and that they are of differing intensities.

Diffraction lines

As can be seen in the diffractogram for calcite (Figure 6.6), several peak or diffraction lines are associated with this crystal. To explain the reason why the diffraction diagram has several peaks, one must consider the structure of the crystal lattice. According to Bragg's law, diffraction will take place between pairs of parallel planes. The angle 2θ on the diffraction diagram is related not only to an integer n (see Figure 6.2) but also to the interplanar distance. It can easily be envisaged that within any regular lattice numerous families of parallel pairs of planes can be identified.

The distance between two parallel atomic planes can be determined from the Miller indices describing the plane. The method of determining the interplanar distance from the Miller index is developed in Appendix 2. To illustrate the relationship between these factors, consider an orthorhombic lattice in which the lattice parameters for a unit cell are a, b and c. Within this simple system, we can consider the number of possible parallel planes that produce diffraction.

It can be shown that for an orthorhombic unit cell the interplanar spacing between any two planes defined by the Miller index (hkl) is given by the equation:

$$\frac{1}{d_{hkl}^2} = \frac{h^2}{a^2} + \frac{k^2}{b^2} + \frac{l^2}{c^2}$$

For a simple cube:

$$\frac{1}{d_{hkl}^2} = \frac{h^2}{a^2} + \frac{k^2}{a^2} + \frac{l^2}{a^2}$$

$$d_{hkl} = \frac{a}{\sqrt{h^2 + k^2 + l^2}}$$

(6.2)

Substituting this equation into the Bragg equation, it can be seen that there must exist a number of diffracted rays corresponding to the various directions of planes in the crystal lattice.

For a simple cubic structure the Bragg equation can be written in the form:

$$\sin^2 \theta = \frac{n^2 \lambda^2}{4a^2}(h^2 + k^2 + l^2)$$

That is, for a lattice structure, families of parallel planes can be identified and characterized by their Miller indices and the Bragg angle can be directly related to these planes.

There will be many possible combinations of parallel planes within the lattice where diffraction may take place (Table 6.1). To illustrate this phenomenon, consider the diffraction diagram for calcium oxide. Calcium oxide has a face-centred cubic structure, i.e. a cube with a node at each corner and a node at the centre of each face. Figure 6.7 and Table 6.1 set out some of the planes that allow diffraction in a face-centred cubic structure. The interplanar distances can be calculated using Equation 6.2 and the diffraction angles calculated using Bragg's law (Equation 6.1). A comparison of this prediction with the observed spectrum for calcium oxide shows a good agreement.

Intensity of a diffraction line
It is evident from the examination of a diffraction diagram that not all parallel planes produce a diffracted beam and the intensity peaks observed are not of the same magnitude. The intensity of these peaks is determined

Table 6.1 *Diffraction angles 2θ are predicted for a simple cubic crystal structure. An interatomic distance (a) of 4.81 Å is taken and the X-ray source is assumed to be Cu Kα − λ = 1.54 Å*

h	k	l	$h^2 + k^2 + l^2$	d (Å)	2θ	θ
1	1	1	3	2.777	32.191	16.10
2	0	0	4	2.405	37.342	18.67
2	2	0	8	1.701	53.839	26.92
3	1	1	11	1.450	64.130	32.06
2	2	2	12	1.389	67.351	33.68

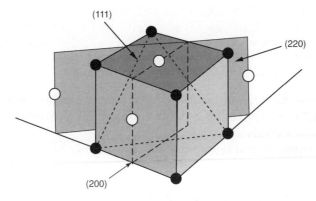

Figure 6.7 *Illustration of the possible atomic planes potentially capable of refraction.*

by the structure of the crystal lattice. The intensity of the diffracted beam is linked to the structure factor F_{hkl}, which is one of the fundamental quantities used in crystallography.

$$F_{hkl} = \sum_{J}\left(f_{y}\ \exp(2\pi i(hx_{J} + ky_{J} + lz_{J})) \right)$$

where F_{hkl} = structure factor, which accounts for the phases of the scattered waves, f_{y} = atomic scattering factor, which accounts for the amplitude of the scattered waves, J = atoms in the lattice at coordinates x_{J}, y_{J} and z_{J}, and hkl = Miller index of the reflecting plane.

Not all the families of planes (hkl) will diffract the incident ray; the structure factor must be non-zero if diffraction is to take place.

Determination of mineralogical composition

To determine the mineralogical composition of a sample, the interplanar distances can be evaluated and compared with those given by standard calibration data. The diffracted beam, at 2θ on the diagram, is observed and the wavelength of the X-ray used is known; an application of Bragg's law (Equation 6.1) enables the calculation of each interplanar distance from the observed values of θ.

Powder diffraction files

The Joint Committee on Powder Diffraction Standards (JCPDS)[7] maintains a comprehensive record of powder diffraction files (PDFs). The original format for the PDF was a 3×5 inch card; as the database developed it became necessary to use computer-based search and matching systems. Since 1987 the production of cards ceased and the data can now be accessed from hardbound book format or computer file. From 1997 the PDF database contained over 77 000 diffraction patterns in 47 data sets; these sub-files include cement materials. In addition to the JCPDS databases, other data more specific to cementitious materials have been drawn up, e.g. cement composition.[8]

The PDF contains the following information:

- a unique PDF identification number
- mineral name
- chemical formula
- crystal unit cell and symmetry
- physical constants
- classification by decreasing interplanar distance (d_{hjk}, intensity and hkl)
- reference to the source of the data
- optical properties.

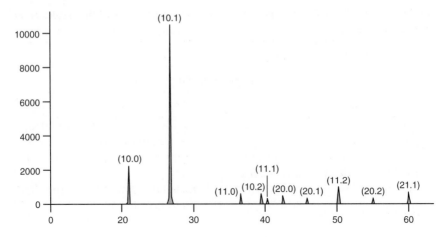

Figure 6.8 *Diffraction diagram of α quartz (filtered Cu Kα). The Miller indices relating to each family of diffraction planes are noted against the peaks.*

The card system also includes a classification of the strongest d_{hkl} in decreasing order to aid identification.

Example

To illustrate the interpretation of an X-ray spectrum consider the diagram shown in Figure 6.8. The diffractometer used X-ray wavelength Cu Kα, that is, a wavelength λ of 1.54 Å. Noting the values of the first four peaks, the interplanar distances can be calculated using Bragg's law (Equation 6.1):

$2\theta°$	d_{hkl} (Å)
20.87	4.25
26.66	3.34
36.52	2.46
39.45	2.28

To identify a mineral an initial diagnosis is made by referring to the most intense peak. The most intense ray is that at 26.66°, i.e. $d_{hjk} = 3.34$ Å. This will yield a group of possible minerals. The process of identification then moves to the second most intense peak, and similarly to the third peak. These observed data can be cross-referenced to the standard reference data in Table 6.2.

The Miller indices for the diffraction planes are shown in Figure 6.8. From an examination of the data, it can be seen that:

- the Miller index of the form (10.0) indicates that a hexagonal Miller–Bravais system is being used
- the interplanar distance decreases as the Miller indices increase.

Table 6.2 *Powder diffraction file for α quartz. The values in bold correspond to the planes indicated in Figure 6.7*

α Quartz: Unit cell   Triagonal $a = 4.913\,\text{Å}$ $c = 5.405\,\text{Å}$

		X-ray powder data			
d (Å)	Strength	*hkl*	*d* (Å)	Strength	*hkl*
4.26	s	**10.0**	2.128	w	20.0
3.343	vvs	**10.1**	1.980	w	20.1
2.458	mw	11.0	1.817	ms	11.2
2.282	w	**10.2**	1.801	vvw	0003
2.237	w	11.1	1.672	w	21.2

s: strong; w: weak; v: very; m: moderately.

General remarks concerning the interpretation of X-ray diagrams

Grain size

A sample of a crystal powder with a size of about 10 μm will produce sharp diffraction peaks. Very small grain sizes will tend to cause a widening of the diffraction peak.

Shape of the trace

- A wide halo beneath the peaks can be attributed to two possible mechanisms; first, a large proportion of amorphous material in the sample and, secondly, X-ray fluorescence. As discussed in Appendix 2, it is possible to achieve a liberation of a photon of X-rays due to the displacement of a core electron (X-ray fluorescence); for example, the *K*α wavelength for copper will create fluorescence in samples containing iron atoms.
- Wide peaks may be an indication of too small a grain size, poor crystallinity or residual stress in the crystallite.

Accuracy of θ and intensity

The peak position may vary slightly at low angles owing to random errors inherent in X-ray analysis.

The intensity of the diffracted peak can vary from the values indicated in the PDF data. Some of the possible reasons for this are:

- the size of crystals in the sample
- preferential orientation: particles with a plate-like structure (such as mica or clays) will tend to take up a parallel orientation in the sample. The effect of this is that one crystal orientation, i.e. one diffraction plane (Miller index), will be dominant in the diffraction diagram
- fluctuations in the performance of the X-ray source and counter.

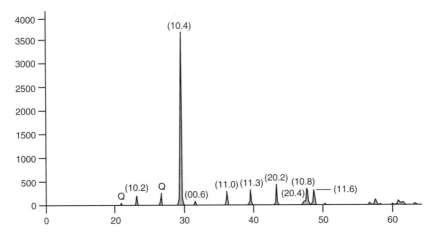

Figure 6.9 *Diffraction diagram for calcite obtained from limestone (Cu Kα filtered X-ray). Calcite has a rhombic form, with the most intense peak corresponding to 3.035 Å (29.4° 2θ). The two peaks marked Q correspond to quartz, which exists as a trace in the limestone sample used in the test.*

Analysis of mixtures

So far, XRD spectrum obtained from a single mineral has been discussed. In the context of analysis of cement-based materials a wide range of crystalline and amorphous compounds will co-exist. To illustrate the effect of a mixture of compounds on the diffraction spectrum a mixture of quartz and calcite is now considered. The diffraction diagram for quartz is shown in Figure 6.8 and that for calcite $CaCO_3$, obtained by grinding a limestone, is shown in Figure 6.9. It can be seen from Figures 6.8 and 6.9 that the peaks are less intense for calcite than for quartz (maximum intensity of calcite = 4000, Figure 6.9; maximum intensity of quartz = 10 000, Figure 6.8).

The identification of the mineral phases present in a mixture is carried out in the manner described above; however, because of the possible presence of several phases, a method of systematic deduction using the most intense peaks is used. The first step is to use the most intense peak for identification with confirmation from the second peak. If the confirmation fails, the third peak is used in combination with the most intense peak. This process continues until a mineral is identified and then all the peaks for that mineral are indexed on the diffraction diagram. The identification then moves to the peaks not indexed and the procedure is repeated. It should be noted that if two minerals have an identical interplanar distance d_{hkl}, the intensity of that peak seen in a mixture will be the sum of the two individual peaks.

It can be seen from Figure 6.9 that after indexing the calcite two peaks remain, the most intense quartz peak being at 3.34 Å (26.7° 2θ). This corresponds to the small quantities of quartz in the limestone source of the sample.

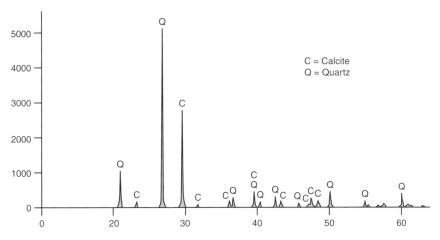

Figure 6.10 *Diffraction diagram for a sandy limestone (Cu Kα filtered X-ray). The sample consists of a mixture of calcite and quartz, the corresponding peaks are marked with C and Q.*

Quantitative XRD analysis

Having identified the compounds present in a mixture, a quantitative analysis may be carried out to determine the relative proportions of the component mineral species. Figure 6.10 shows the diffraction diagram for a sample of sandy limestone; this naturally occurring rock is a mixture quartz and calcite, and the relative proportion of these two minerals can be estimated from this diagram. The two principal peaks at 26.7° and 29.4° 2θ correspond to the most intense peaks for quartz and calcite, respectively. The quartz intensity is reduced, 10 000 for quartz alone (Figure 6.8) and 5000 for the sandy limestone (Figure 6.10). Information concerning intensity can be used to make a quantitative analysis of a mixture. A relationship exists, not necessarily linear, between the intensity of diffracted rays and their relative proportional quantity in a mixture. To carry out a quantitative study a calibration must be undertaken. A series of samples must be prepared containing known quantities of the minerals. Analysis of these calibration mixtures will enable the intensity of a particular peak to be plotted against the proportion of that mineral in the mixture. Figure 6.11 shows a calibration curve for mixtures of calcite and quartz.

Returning to the diffraction diagram for the sandy limestone, it can be seen that the intensity of the quartz peak 3.34 Å (26.7° 2θ) is 5200. Using the calibration curve shown in Figure 6.11, the percentage of quartz can be estimated either graphically or from the regression equation. Thus, it is deduced that the percentage of quartz in the sandy limestone is 54.4%.

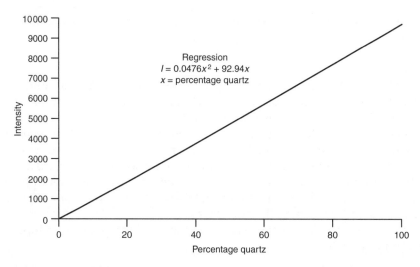

Figure 6.11 *Calibration curve for a mixture of quartz and calcite. The reference peak is taken as the quartz peak 3.34 Å (26.7° 2θ).*

Internal standard method

The internal standard method (sometimes called the spiking method) is a technique for determining the quantity of a certain crystal phase by measuring peak heights. The absolute values of peak heights are variable; therefore a crystal, the internal standard, is added to the sample to act as a reference.

The method is based on the precept that the intensity of a diffraction line i is $I_{i\alpha}$ from a phase α and is related to the proportional weight of that phase X_α in a mixture:

$$I_{i\alpha} = \frac{K X_\alpha}{\rho_\alpha \left(\dfrac{\mu}{\rho}\right)}$$

where K is a constant, ρ_α is the density of the phase α and (μ/ρ) is the mass absorption coefficient for the mixture. The problem in applying this approach is in the determination of the mass absorption coefficient: this term reflects the absorption of X-rays during elastic or inelastic interactions between all the elements in the mixture and therefore its value will be unique to any mixture. The internal standard method is based on the comparison of two diffraction lines, one for the phase to be quantified, α, and the other for the internal standard s, the internal standard being a crystal phase that does not occur in the original sample.

Taking the ratio of the two intensities, the equation can be simplified:

$$\frac{I_\alpha}{I_s} = k\frac{X_\alpha}{X_s}$$

That is, the intensity of the peak is proportional to its relative weight. The quantity phase α in the sample can be varied and the intensity of the peak referenced to that of the internal standard.

Boehmite Al O(OH), for example, has been proposed as a suitable internal standard for the quantitative XRD of certain sulfate minerals in concrete.[9] This method requires that reference samples are made up with a fixed quantity of the internal reference and variable quantities of the target phase. Calibration curves of the type shown in Figure 6.12 are produced which can then be used for the quantitative analysis.

The following considerations must be taken into account when adopting quantitative XRD.

- The calibration curves are unique to a specific sample.
- The internal reference must have peaks, which are close to but do not overlap the main peaks of the mineral being studied.
- The internal standard must have as few rays as possible. Cubic structured crystal forms are preferential, such as CaF_2 and MgO.
- Certain phases encountered in cementitious materials can co-exist as amorphous and crystalline forms. The analysis is therefore limited to a quantitative study of crystal forms.

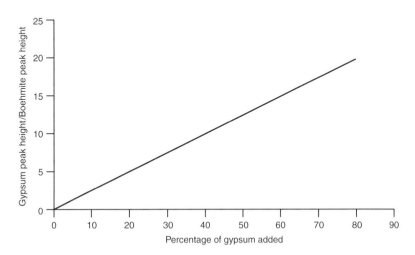

Figure 6.12 *Calibration curve for gypsum, after Crammond.[9] The sample consists of a concrete sample, free of sulfate minerals, to which is added 10% Boehmite and a variable quantity of gypsum. The peaks referenced are Boehmite 14.28° 2θ and gypsum 11.69° 2θ (Cu Kα filtered X-ray).*

Bibliography

Azaraff L.V. and Buerger L.E. The powder method in X-ray crystallography. McGraw-Hill, 1958.

Guinier A. *Théorie et technique de la radiocristallographie*. Dunod, Paris, 1963.

Kittel Ch. *Physique de l'état solide*. Dunod, Paris, 1983.

Klug H.P. and Aleaxander L.E. *X-ray diffraction procedures*. Wiley, 1967.

References

1   Askeland D.R. *Science and engineering of materials*, 3rd edn. Chapman and Hall, London, 1996, pp. 67–78.

2   Frey K. and Kinneging B. X-ray analysis: a review. *World Cement* 2001, **32** (2), 29–30, 32.

3   Schmidt R. and Kern A. Quantitative XRD phase analysis. *World Cement* 2001, **32** (2), 35–36, 38, 41–42.

4   Voinovitch I.A. and Louvrier J. L'analyse et l'identification des constituants. In *Le béton hydraulique*. Presses de l'École Nationale des Ponts et Chaussées, Paris, 1982, pp. 83–94.

5   Eberhart J.-P. *Analyse structurale et chimique des matériaux*. Dunod, Paris, 1997.

6   Milburn G.H.W. *X-ray crystallography*. Butterworth, London, 1973.

7   Joint Committee on Powder Diffraction Standard. *Powder diffraction files*. International Centre for Diffraction Data, Swarthmore, PA.

8   Taylor H.F.W. Appendix: Calculated X-ray powder diffraction patterns for tricalcium silicate and clinker phases. In *Cement chemistry*, 2nd edn, ed. H.F.W. Taylor. Thomas Telford, London, 1997, pp. 385–391.

9   Crammond N.J. Quantitative X-ray diffraction analysis of ettringite, thaumasite and gypsum in concretes and mortars. *Cement and Concrete Research* 1985, **15**, 431–441.

Chapter 7

Scanning electron microscopy and micro-analysis

For background reading, see references 1–3 and Appendix 2.

Scanning electron microscopy (SEM) (see Figure 7.1), is now a well-established method that can offer useful information concerning the structure of a material. A range of techniques has been developed to enable a chemical analysis of a material to be carried out in tandem with its visual

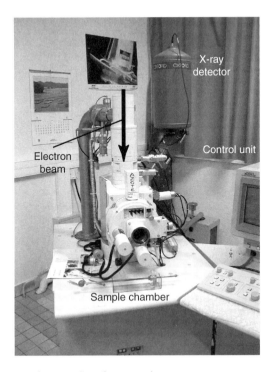

Figure 7.1 *Layout of a scanning electron microscope.*

representation. The physical processes involved in these techniques are described in Appendix 2.

In SEM and micro-analysis a high-energy beam of electrons ~60 keV is directed at a sample and sensors are used to detect the various energy emissions (Figure 7.2):

- **secondary electrons:** result from inelastic interaction between the electron beams and atoms. These electrons have low energy, typically 50 eV, and are emitted from the surface of the sample
- **backscattered electrons:** result from the elastic interaction between the electron beam and atoms. Consequently, the energy of back-scattered electrons is of the order of the transmitted beam 30 keV and the angle of scattering has $2\theta > 90°$
- **X-ray fluorescence:** results from the inelastic scattering from atoms ionized by incident electrons at a depth within the material
- **Auger electrons:** an electron emission due to an inelastic interaction close to the surface of the material
- **transmitted electrons:** in thin samples electrons pass through the sample
- **absorbed electrons:** may be absorbed by the sample.

The type and intensity of emission will be dependent on the atomic number of the atom, sample thickness and the energy level of the incident electron beam. In the following sections three techniques will be discussed: topographical mapping using secondary electrons, topographical and

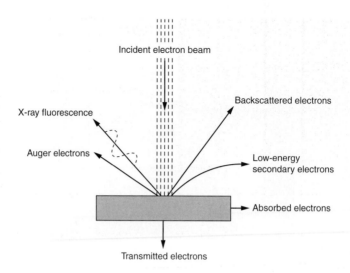

Figure 7.2 *Emissions from a sample subjected to an electron beam.*

elemental mapping using backscattered electrons, and elemental analysis using X-ray fluorescence. Before discussing the techniques, it is of interest to consider the origin of the various emissions. Figure 7.3 illustrates the zone from which these emissions emanate.

Figure 7.3 shows that the resolution of an image, the width of the zone affected, varies with the technique used. The resolutions are depicted by the dimensions S, B and X, corresponding to the three emissions secondary, backscatter and X-ray. These resolutions are of the order of $S \sim 50 \text{Å}$, $B \sim 500 \text{Å}$ and $S \sim 1 \mu\text{m}$. It can therefore be seen that the highest resolution image is derived from a detection of the secondary electron emissions.

Scanning electron microscopy

The SEM is capable of producing topographical surface images of a material with magnifications ranging from 30 to 300 000 with a resolution of the order of $30–50 \text{Å}$. In the SEM a focused electron beam about 30 keV and $20–30 \text{Å}$ in diameter is rastered (scanned) across the surface of the sample. The beam excites atoms on the surface of the material and produces an emission of secondary electrons. These electrons are attracted to the detector owing to the low energy of the electron and the high potential at the detector (Figure 7.4).

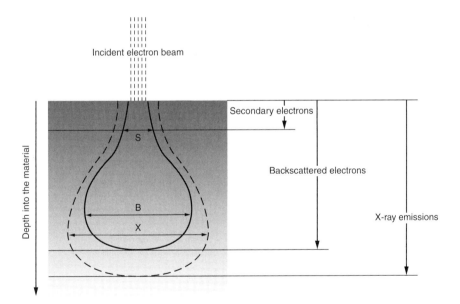

Figure 7.3 *Zones from which emissions occur. The secondary electron emission comes from depths of about 10–40 Å. The backscattered electrons have a deeper origin, up to 1000 Å. The X-ray fluorescence produced in a material covers a much wider zone, delineated by the dashed line, and penetrates to a depth of the order of 1 μm.*

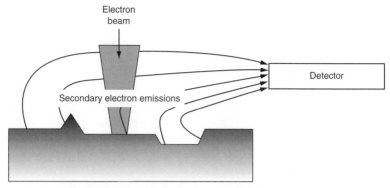

Figure 7.4 *Secondary electron emission and transmission to a detector.*

Figure 7.5 *Sample of material mounted for SEM observation. The sample is coated with a fine layer of gold and electrically connected to the brass support by a conductive strip.*

Thus, it is possible to gain an image of the topography of a material in holes and behind corners. The images are analogous to viewing a landscape in daylight, with areas in shade being discernible.

The detected secondary electron emission forms an image on a cathode ray oscilloscope (CRO) by mapping the intensity of the detected signal against position. The intensity of the secondary electron emission is a function of the angle between the incident ray and the surface of the sample; thus, the intensity of the signal is a function of the topography of the surface. In the case of non-conductive materials, the sample may undergo an accumulation of surface charge; therefore, such materials are coated with a conductive material such as gold and fixed to a brass support with graphite-coated adhesive tape (Figure 7.5).

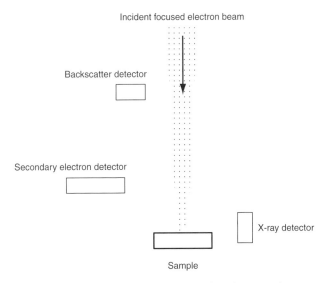

Incident focused electron beam

Backscatter detector

Secondary electron detector

X-ray detector

Sample

Figure 7.6 *Layout of the detectors used in a scanning electron microscope.*

In addition to the topographical representation of the surface, it is possible to detect other emissions from the sample, X-ray, backscattered electrons, etc., to gain information concerning the chemical composition of the sample. Figure 7.6 illustrates the layout of detectors used in the SEM.

Backscatter images

A second mode of assembling an image of a surface is to detect backscattered electrons. The backscattering of electrons from within a crystal lattice is an elastic interaction; therefore, the energy of the emitted electrons is close to that of the transmitted beam. Because of the high energy of the emitted electrons they are transmitted in a rectilinear fashion and not, as in the case of secondary electrons, attracted to the detector. Figure 7.7 illustrates the trajectories of backscattered electrons captured by the detector. Mapping of a surface in this mode will produce images that have a high contrast, and areas in shadow will be totally obscured.

The intensity of the emitted backscattered signal is related to the atomic number of the atoms provoking the scattering. The implication of this is that the contrast seen on a backscatter image can, with suitable calibration, be used to give an element mapping of the surface. The emission of the backscattered electrons originates from below the surface, giving the advantage that non-representative materials polluting the surface will not influence the analysis.

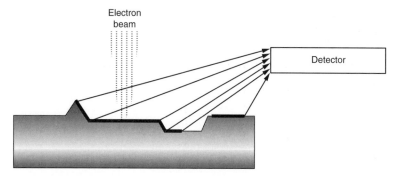

Figure 7.7 *Backscattered electron emission and transmission to a detector. The linear trajectory of the high-energy electrons only permits imaging of the parts of the surface marked with the heavy line.*

Energy dispersive X-ray spectroscopy (EDS or EDX)

EDS is based on the detection of the X-ray emissions from the sample (Appendix 2). The mechanism involved in the emission of X-ray due to the inelastic interaction between electrons and an atom is described in Appendix 2. From this description it can be seen that a photon of X-rays or an Auger electron is released due to the displacement of an electron in one of the orbits (Figure 7.8). The wavelength, and therefore the energy, of the emitted X-ray is a unique characteristic of the element and therefore can be used to make a quantitative and qualitative analysis of the sample. The characteristic of the X-ray photon emitted for an atom will depend on the transition that takes place in the electron structure, in simple terms, $K\alpha$, $K\beta$, $L\alpha$, etc.

The most useful transitions in X-ray spectroscopy are those involving the displacement of one of the core electrons, because the energy levels of the inner electrons are not affected by chemical bonding of the atoms.

Energy-dispersive X-ray detectors

The basic layout of an energy-dispersive X-ray detector is shown in Figure 7.9. The X-ray enters the detector through a beryllium window. The detector uses a photoelectric effect to convert the X-ray emission to an electrical pulse. For energies up to about 30 keV silicon lithium semi-conductors are commonly used; this corresponds to the $K\alpha$ peak for $Z < 55$ (above this atomic number germanium semi-conductors are more efficient). After amplification these pulses are processed by a multi-channel analyser. Since the voltage of the pulse is proportional to the energy of the X-ray it is possible to assemble a spectrum of the number of pulses (counts) with respect to X-ray energy. Figure 7.10 shows a typical SEM image and the associated EDS plot.

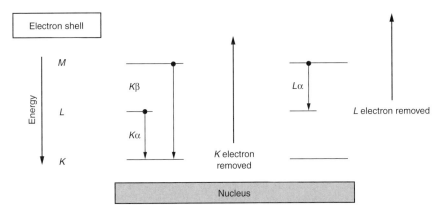

Figure 7.8 *Simplified illustration of the energy level diagram showing possible transitions.*

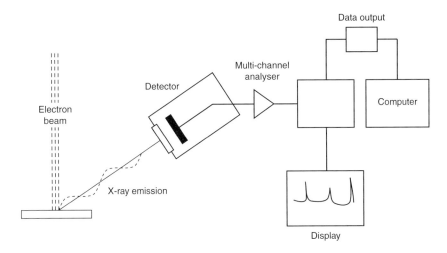

Figure 7.9 *Layout of an energy-dispersive electron analysis system.*

Energy-dispersive spectrum

In the analysis of gypsum $CaSO_4 \cdot 2H_2O$ shown in Figure 7.10, the peaks anticipated for the principal spectral lines for each element are shown in Table 7.1. The wavelength of each peak is converted to energy using the relationship:

$$E = hv = \frac{hc}{\lambda}$$

where h = Planck's constant (6.62×10^{-34} J·s or 4.14×10^{-15} eV·s) and c = velocity of light (3×10^8 m/s).

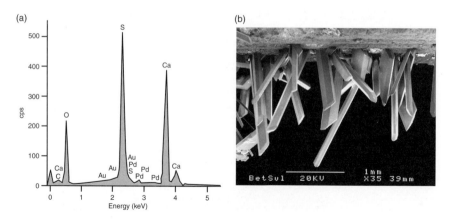

Figure 7.10 *SEM micrograph of gypsum (×16) formed on the surface of concrete due to ammonium sulfate attack. The EDS plot, taken at the surface of a crystal, shows the presence of calcium, sulfur and oxygen, thus confirming the crystal as being gypsum. Close inspection of the energy spectrum reveals that there is more than one peak that relates to calcium: these peaks correspond to the Kα Kβ, etc., wavelengths. Au (gold) and Pd (palladium) corresponds to the gold coating used in the preparation of the sample.*

Table 7.1 *Wavelength and energy levels of X-ray emissions from gypsum*

	Kα		Kβ		Lα	
	λ (Å)	Energy (keV)	λ (Å)	Energy (keV)	λ (Å)	Energy (keV)
Ca	3.36	3.69	3.09	4.01	36.33	0.341
S	5.37	2.31	5.032	2.46	–	–

This rather crude analysis explains the reason for the multiple peaks associated with the energy spectrum obtained in EDS. In reality this analysis is built into the software used in the system and the plotted spectrum is annotated as shown in Figure 7.10. An EDS analysis of data will also include an elemental and atomic analysis. Table 7.2 sets out the data recorded for gypsum.

In this type of output the identity of a species can be confirmed. The column 'atomic' in Table 7.2 gives an indication of the number of atoms in the species being examined.

To illustrate this analysis, the equation for gypsum is $CaSO_4 \cdot 2H_2O$. The number of atoms is counted and percentages are ascribed, noting that hydrogen is not detectable by EDS (Table 7.3). This analysis shows that there is a reasonable agreement between the expected and observed distribution of atoms.

Table 7.2 *Analysis of the gypsum crystals shown in Figure 7.10*

Element	Element %	Atomic %
Ca	25.85	14.05
S	22.03	14.97
O	52.13	70.98

Table 7.3 *Expected and observed distribution of atoms in gypsum*

Element	No. of atoms	Atomic % (theory)	Atomic % (observed)
Ca	1	1/8 = 12.5%	14.1%
S	1	1/8 = 12.5%	15%
O	6	6/8 = 75%	71%
H	4	Not registered	

Table 7.4 *Summary of analytical techniques*

Scanning electron microscopy (SEM)	The SEM can produce images from secondary electron emission (high resolution) or from backscattered electrons (lower resolution), but contrast is a function of atomic number. *Practical considerations:* • Non-conductive samples must be coated with a conductive film • The sample must be compatible with a vacuum (with special sample preparation techniques the observation of organic cells is possible) • Size limitations: diameter less than 100 mm, height less than 50 mm • The depth of field allows the topographical mapping of rough surfaces • Good surface resolution: $\sim$30 Å • Hidden points and interior of holes in the surface can be viewed by secondary electrons
Energy dispersive X-ray spectrometry (EDS or EDX) Electron probe micro-analysis (EPMA)	The technique can either be applied for point analysis or produce scans of a surface to map specific elements. *Practical considerations:* • Non-conductive samples must be coated with a conductive film • The sample must be compatible with a vacuum • Size limitations: diameter less than 100 mm, height less than 50 mm

Table 7.4   *(Continued)*

	• Rapid chemical analysis of areas of a solid (typically 0.5–3 μm diameter) • Sampling depth 1–2 μm • For polished surfaces the composition of the specimen can be mapped • Profiling of rough surfaces is not possible • Quantitative analysis is possible if an internal reference is present • Detection limit, typically 0.1%, dependent on the element • Element range detectable: carbon to uranium, $Z = 6$–92 • The material must have a flat surface: emissions from non-horizontal surfaces may not be trapped by the detector
Auger electron spectrometry (AES) Scanning Auger microscopy (SAM)	A focused electron beam (3–25 keV) is used to irradiate an area of the sample. Elements have unique Auger electron's characteristics. The detection and analysis of these electrons enables the chemical analysis. This technique is used for analysis of surfaces and thin films. *Practical considerations:* • The sample must be conductive • The maximum sample size is of the order of 25 mm diameter and 10 mm thick • The sample must be vacuum compatible; needs a high vacuum • Capable of depth profiling to a depth of 10 μm • Resolution: surface 150 Å, depth 20 Å • The technique favours the detection of lighter elements
X-ray fluorescence (XRF)	Inelastic interaction of an X-ray beam producing characteristic X-ray emissions from the element in the sample. The degree of X-ray fluorescence of an element is dependent on the wavelength of the incident ray; it is therefore necessary to select an X-ray source suitable for the elements being sought. The intensity and energy of the emissions are measured using a Li(Si) detector and enable the identification of elements in the sample. *Practical considerations:* • Elements range detectable: sodium to uranium, $Z = 11$–92 • Very good detection limits

Table 7.4 (*Continued*)

Electron backscattered diffraction (EBSD)	Electron backscattered diffraction was an effect discovered by Kikuchi in 1928. Backscattered electrons are produced by the elastic scattering of electrons at a depth of a few nanometres within a crystal matrix. As the electrons pass through the crystal lattice, diffraction takes place according to Bragg's law. The emitted image, comprising two lines, is projected onto a phosphorous screen. The Bragg angle, and therefore interplanar spacing, can be deduced

This discussion of micro-analysis has concentrated on the use of EDS, which is commonly used for a rapid analysis of a material. Used in conjunction with SEM it enables a rapid elemental analysis of crystal and amorphous forms found in a sample. EDS can be used in two modes. First, as seen in Figure 7.10, it can be used to make a point analysis, which is very useful in the identification of crystal forms seen on an SEM micrograph. The second mode uses the scanning facility used in the production of an SEM image. Using this facility it is possible to map a surface of a material for a specific element. The superposition of these element maps with the SEM image enables an interpretation of the elemental composition of the sample in a global sense. Similarly, the SEM linked to EDS can produce a profile of elements across a section of the material.

A summary of analytical techniques is given in Table 7.4.

References

1 Askeland D.R. *Science and engineering of materials*, 3rd edn. Chapman and Hall, London, 1996, pp. 67–78.
2 Voinovitch I.A. and Louvrier J. L'analyse et l'identification des constituants. In *Le béton hydraulique*. Presses de l'École Nationale des Ponts et Chaussées, Paris, 1982, pp. 83–94.
3 Eberhart J.-P. *Analyse structurale et chimique des matériaux*. Dunod, Paris, 1997.

Chapter 8

Physicochemical examination of concrete

This chapter develops the application of the techniques discussed previously with the aim of describing cements and concrete. It is essential to obtain an overview of the general structure of concrete before examining the deteriorated material. Most of the analyses discussed are based on scanning electron microscopy (SEM) and X-ray diffraction (XRD), but these techniques are often applied to complement or extend the initial petrographic methods.

One of the main problems with SEM and XRD is the size of the sample; both techniques can only examine very small sample sizes, of the order of a few grams. More global application of these techniques will inevitably lead to problems of representative sampling and uncertainty in the interpretation of results. It is therefore vital that the samples are representative. If problem areas have been identified by optical methods, it is relatively simple to pinpoint small areas for detailed study. However, the need to adhere to good representative sampling techniques cannot be ignored. SEM and XRD can contribute to the analysis of concretes in many applications, including:

- determining cement composition
- detecting conversion of high alumina cement (HAC)
- compositional mapping and profiling of concrete sections
- assessing the degree of hydration of cements and pozzolans
- estimating the water/cement (w/c) ratio
- examining cement–aggregates interfaces
- identifying physicochemical changes due to biochemical attack
- studying sulfate attack: thaumasite and delayed ettringite formation (DEF)
- identifying alkali aggregate reaction (AAR)
- detecting carbonation
- mapping chloride concentrations
- assessing changes due to fire damage.

Some of these applications will be discussed in this chapter and those orientated towards deteriorated concrete will be discussed in the case histories (Chapter 9).

Determination of the composition of cement

X-ray diffraction

Ordinary Portland cements (OPC) are crystalline in nature and therefore XRD methods could be used to determine the relative composition of the basic compounds. Figure 8.1 shows the X-ray diffractogram for an OPC. What is immediately apparent from this figure is the superposition of several of the principal peaks for the main phases. This superposition poses severe problems for the application of quantitative XRD. In addition, cement contains a glass (amorphous) phase, which will not be apparent in XRD analysis. Because of the complexity of the diffractogram for a cement, the application of quantitative XRD analysis in the determination of cement composition is rather restrictive. Recent developments in computer programs for the interpretation of XRD analysis make direct interpretation of the crystal structure possible.

Standard test methods exist (ASTM C1365-98[1]) for the quantitative XRD of certain phases of Portland cement clinkers: tricalcium aluminate, ferrite

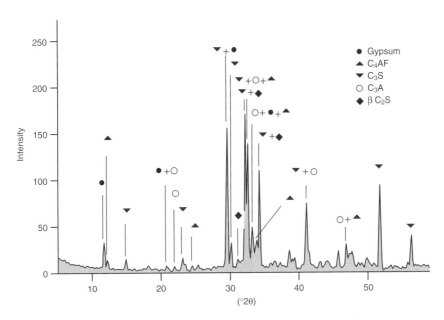

Figure 8.1 *X-ray diffractogram for an ordinary Portland cement (Kα Cu filtered X-ray). The strongest peaks, at 28–35°, are mostly combinations of phases.*

and periclase. Commercial methods have been proposed for the determination of various phases, such as free lime. In this case the principal peak for CaO (CaO 2.638 Å, 33.9° 2θ; C_3S 2.776 Å, 32.21° 2θ and 2.602 Å, 34.43° 2θ) overlaps with one of the calcium silicate peaks and therefore the analysis has to be based on a secondary peak.

X-ray fluorescence

One approach to determining cement composition is to carry out an oxide analysis and calculate the composition. The oxide analysis of the cement was traditionally carried out by chemical analysis; however, an elemental analysis of a cement can be determined from the X-ray fluorescence of a cement. This can be carried out either by X-ray fluorescence analysis or from an energy dispersive X-ray spectrometry (EDS) scan of a sample of cement. Interpretation of the results of this elemental composition will require an analysis based on calibration studies.

The equations proposed by Bogue[2] are used to determine the compound composition of cement from the oxide analysis. He termed these composition values potential composition, because the equations are based on the assumption that the clinker is a fully crystallized material, which would imply very slow cooling. Bogue equations for ordinary Portland cements are:

$$C_3S = 4.0710C - 7.6024S - 1.4297F - 6.7187A$$
$$C_2S = 8.6024S + 1.0785F + 5.0683A - 3.0710C$$
$$C_3A = 2.6504A - 1.6920F$$
$$C_4AF = 3.0432F$$

where C, S, A and F represent the proportional weight of the oxides CaO, SiO_2, Al_2O_3 and Fe_2O_3.

In reality, rapid cooling gives rise to an amorphous glass phase which distorts the values of the composition given by the Bogue equations. These equations tend to overestimate the alite content and underestimate the C_3A content. The stoichiometry of the ferrite phase is variable depending on the aluminium and iron contents. The composition can vary between C_6A_2F and C_6AF_2. For many OPCs the composition is close to C_4AF; however, for low C_3A contents the composition is closer to C_6AF_2. In addition, an increased iron content has the effect of reducing the reactivity of the ferrite phase.

Morphology of hydration products

To help the reader with the interpretation of SEM micrographs a brief morphology of the principal phases found in hydrated cement paste are presented here. The principal XRD peaks are set out in Appendix 4.

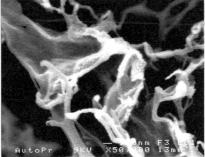

Figure 8.2 *CSH(II) in ordinary Portland cement, CPA CEM I (left ×2500; right ×25 000).*

Portland cements

CSH

Structurally, CSH can be considered as an amorphous material, i.e. a weak regular atomic arrangement, which has a layered structure similar to that of a clay. Figure 8.2 shows that CSH(II) has an irregular structure and a high porosity. The structure of CSH varies with its degree of maturity. Its initial structure consists of a needle-like gel that develops into a honeycomb structure. As curing continues, the structure becomes featureless. The structure is more crystalline in steam-cured concrete, as is also seen in the case of some fire-damaged concretes. The sheets are bonded by a weak hydrogen bond with water between the plates or, in the case of dehydration, the plates collapse and a stronger Si–O–Si bond is formed. This is one reason why dry concretes are stronger than saturated concretes.

On XRD analysis, CSH(I) or CSH(II) formed under normal curing temperature shows weak and diffuse peaks, whereas steam-cured or fire-damaged concrete will show much sharper peaks. Concretes cured at elevated temperature from other calcium silicate hydrates, such as 11 Å tobermorite and 9 Å tobermorite, are more crystalline than CSH and can be identified by XRD.

Portlandite

Calcium hydroxide, $Ca(OH)_2$; structure: hexagonal plates (00.1) cleavage.

In contrast to CSH, Portlandite is crystalline, having hexagonal plates with strong cleavage planes (Figure 8.3). Portlandite is deposited within the CSH and within the void structure of the cement matrix. It blocks pores and capillaries and therefore reduces porosity and permeability. This species is soluble and reactive with other chemical species.

Figure 8.3   *Portlandite formations in the cement paste (left ×500; right ×1200).*

Figure 8.4   *Ettringite formations in a Portland cement (×1000).*

Ettringite

Hydrated calcium aluminium trisulfate, $Ca_6Al_2(SO_4)_3(OH)_{12} \cdot 26H_2O$; structure: hexagonal.

Ettringite crystals have an acicular, needle-like structure a few micrometres in cross-section (Figure 8.4). Above 60°C ettringite starts to convert to monosulfate and above 100°C it is completely dehydrated. The primary ettringite formed in hydration will convert to the monosulfate in the presence of C_3A.

Monosulfate

Tricalcium aluminate monosulfate 12-hydrate, $3CaO \cdot Al_2O_3 \cdot CaSO_4 \cdot 12H_2O$; structure: hexagonal plates (00.1) cleavage.

These crystals occur as very small formations (Figure 8.5). The monosulfate is normally the product of ettringite and C_3A. It will convert back to ettringite in the presence of sulfates.

Figure 8.5 *Monosulfate formation in Portland cement (×2500).*

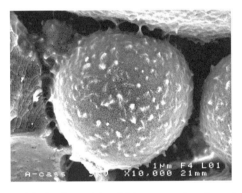

Figure 8.6 *Pulverized fuel ash particle within a cement paste matrix. Note the surface growth of hydration products (×5000).*

Blended cements

Slag and pulverized fuel ash (PFA) pozzolans consist predominantly of silicate glass. PFA is normally identifiable as spherical grains as small as 5 μm, whereas slag particles tend to be more angular in nature. In hydrated cement paste these pozzolans can remain unaffected for several years. In both cases the particles commonly have Portlandite and CSH forms growing from the surface.

Concretes containing PFA OPC blended cements tend to have a characteristic dark appearance. PFA particles consist of aluminosilicate glass, with crystalline inclusions of mullite ($Al_6Si_2O_{13}$) and quartz; there is frequently a surface coating rich in magnetite and haematite (iron oxides) derived from the combustion of pyrite. Figure 8.6 shows a PFA particle with the surface growth of CSH and Portlandite.

When blast furnace slag is cooled rapidly it forms a large proportion of calcium–magnesium–silicate glass, typically 90%. The admixture ground-

Figure 8.7 *Grains of ground-granular blast furnace slag; the growth of CSH can be seen at the periphery of the grains (×500).*

granular blast furnace slag (GGBFS) gives the concrete a greenish blue colour. Within the amorphous glass phase there are various crystalline phases containing merwinite or melilite. The proportion of crystalline to glass phases in GGBFS has been proposed as a method for determining the reactivity of the slag.[3] Figure 8.7 shows blast furnace slag grains in a mortar mix.

Studies on the hydration of cements

Two studies are described that were carried out at INSA Rennes to examine the influence of hydration temperature on cement mortars.

Ettringite formation in OPC

The first experiment was carried out on a standard OPC mortar. The hydration temperatures were 20°C to simulate normal conditions, and 70°C to simulate steam cure. At the end of the hydration the samples were examined by XRD analysis using a diffractometer with a copper *K*α X-ray source. The diffractograms for the analysis are shown in Figure 8.8. It can be seen from the 20°C diffractogram that only ettringite exists; this also shows that at this age the cement paste has not begun the conversion from ettringite to monosulfate. The 70°C diffractogram shows that monosulfate has formed in preference to ettringite.

The hydration reaction is accelerated at a higher temperature and therefore there is a rapid increase in strength. For this reason steam curing of concrete is often used in pre-cast production to speed up production. At high curing temperatures, there is a suppression in the formation of ettringite (as shown in this study). Concrete cured in this way can develop DEF if exposed to moist conditions.

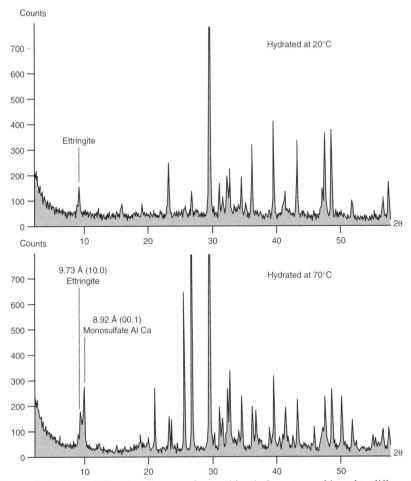

Figure 8.8 *X-ray diffraction diagrams for two identical mortars subjected to different hydration temperatures. The extracts of the diagrams show the peaks corresponding to ettringite 9.73 Å (9.07° 2θ) and monosulfate 8.92 Å (9.9° 2θ). Upper diagram: with curing at 20°C only an ettringite peak occurs; lower diagram: curing at 70°C shows a dominant presence of the monosulfate.*

Hydration of ciment fondu

The second study was carried out on mortar made with ciment fondu. To illustrate the hydration process the change in morphology was studied using XRD. Figure 8.9 shows the diffractogram for ciment fondu. The principal components are monocalcium aluminate (CA), gehlenite (C_2AS) and calcium disilicate (βC_2S).

The effect of temperature on the hydration curing process was studied. As in the previous case the temperatures used were 20°C and 60°C. The diffractograms from the two specimens are shown in Figure 8.10.

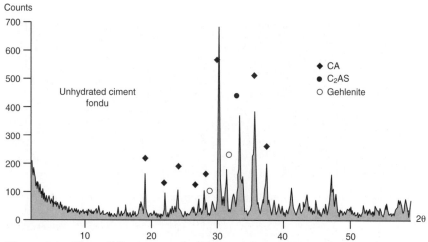

Figure 8.9 *X-ray diffractogram of unhydrated ciment fondu. The principal phases are monocalcium aluminate (CA), gehlenite (C_2AS) and calcium disilicate (βC_2S).*

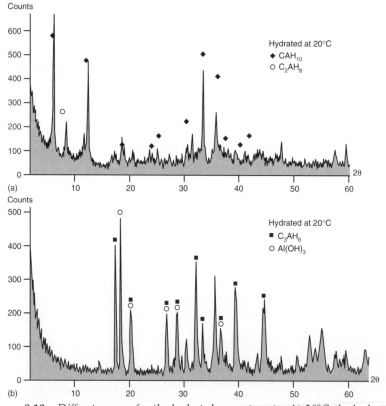

Figure 8.10 *Diffractograms for the hydrated cement paste. At 20°C the hydration products are monocalcium aluminate (CAH_{10}) and a lesser quantity of dicalcium aluminate (C_2AH_8), this being a common hydration reaction for alumina cements. At 60°C the hydration produces hydrogarnet (C_3AH_6) and gibbsite (AH_3).*

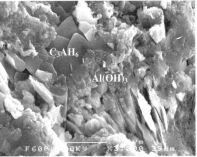

Figure 8.11 *Effect of curing temperature on the hydration of ciment fondu. Left: cement paste after curing at 20°C (×9000); right: sample cured at 60°C (×1500). CAH_{10} has a hexagonal prism structure, C_3AH_6 has a cubic or trapezohedral structure, and gibbsite ($Al(OH)_3$) has a monoclinic structure with (001) cleavage.*

The normal initial hydration of a high alumina cement results in the production of CAH_{10} and a smaller quantity of C_2AH_8, which is in agreement with the diffractogram shown in Figure 8.10. These calcium aluminate hydrates tend to be unstable and can undergo a conversion to cubic hydrogarnet C_3AH_6, a monoclinic gibbsite. Micrographs of the two hydrated concretes are shown in Figure 8.11. The mineral forms have been identified by EDS.

This conversion reaction results in an increase in porosity and a reduction in strength. This study shows that hydration and curing at the higher temperature provoked the conversion at the onset of the concrete's life.

Study on hardened concrete

Samples

Two concrete samples were examined in this study. The concrete was approximately 10 years old; it was produced under laboratory conditions, cured for 7 days under water and then stored in the laboratory. These samples had not been subjected to an environment more severe than normal atmosphere. OPC was used in the fabrication of the specimens. One sample was made with a flint gravel coarse aggregate and a siliceous sand, while the second concrete, a lightweight concrete, contained 'lytag', a sintered PFA, coarse aggregate and siliceous sand.

Figure 8.12 shows a polished surface of the samples. From a cursory visual inspection of the samples it can be seen that:

- there is no major cracking in either sample
- the aggregate is well and uniformly distributed through the sample
- the cement paste has no sign of softening or deterioration.

 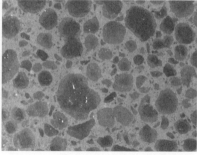

Figure 8.12 *Surface of two concrete samples (5 × 5 cm². Left: conventional OPC concrete with a flint gravel coarse aggregate; right: lightweight concrete with lytag coarse aggregate.*

The lytag aggregate has a two-tone appearance, a rust-coloured outer zone shell and a dark, almost black, core.

Small cores, 15 mm in diameter, were taken from the samples and prepared for SEM analysis. The initial phase of this study comprised surface images with secondary and backscattered electron emissions.

A section of the sample with flint gravel aggregate was selected and examined by compositional mapping, and finally the cement paste and aggregate cement paste were examined in detail.

Low-magnification observation

The surface was initially observed at a magnification of ×25. The secondary electron and backscatter images of the two samples are shown in Figure 8.13. In comparing the ×25 magnification images the difference in resolution can be clearly seen, the secondary electron image having the best resolution. The backscatter image is of value in considering the elemental composition of the samples.

Concrete with flint gravel aggregate

On the backscatter micrographs the changes in grey scale between particles indicate different elemental compositions. Four main shades appear on the backscatter images. Various particles can be identified against a general mid-grey background:

A   the darkest appear to be the coarse and fine aggregates
B   scattered through the micrograph there are light, also most white, grains
C   background
D   there is one particle at the centre of the micrograph with a tone lighter than the background.

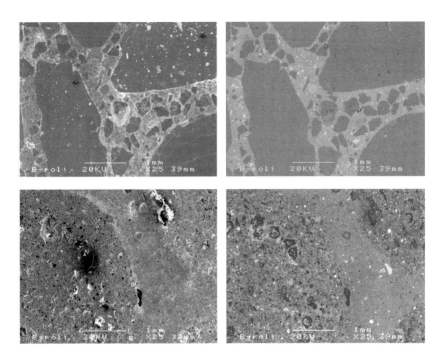

Figure 8.13 *Micrographs (×12) of concrete with flint gravel aggregate (top) and with lytag aggregate (bottom). Left: micrographs obtained from secondary electron emission; right: backscatter images of the same area.*

To identify these features the EDS function on the SEM is used. Table 8.1 shows the atomic percentage for each of these areas together with a preliminary identification.

- **Area A:** corresponds to the aggregate, testing the hypothesis that the aggregate is a silicate, SiO_2. This corresponds to 33.3% Si and 66.7% O; there is a reasonable agreement with the observation.
- **Area B:** the light-coloured grains. The granular nature plus the elemental composition of calcium silica and oxygen indicate strongly that this could be unhydrated cement:

    C_3S, $3(CaO) \cdot SiO_2$; 3 Ca, 1 Si and 5 O, i.e. 33% Ca, 11% Si and 55% O
    C_2S, $2(CaO) \cdot SiO_2$; 2 Ca, 1 Si and 4 O, i.e. 29% Ca, 14% Si and 57% O

    The analysis indicates a marginal preference for C_2S.
- **Area C:** the background area. This is probably CSH. The form of CSH(II) is variable, $2CaO \cdot SiO_2 \cdot nH_2O$ ($n = 2-4$), implying 2 Ca, 1 Si and 7 O (assuming $n = 3$), i.e. 20% Ca, 10% Si and 70% O. The percentage of calcium atoms is greater than expected; this can be explained by the fact that CSH and Portlandite co-exist (see Figure 8.3) and that the sampling zone of EDS is relatively large. Given the

Table 8.1 *Atomic percentages of areas A–D*

	A	B	C	D
O	64.8	58.4	75.5	64.4
Mg				2.1
Al				2.5
Si	35.2	10.7	6.5	17.0
K				3.4
Ca		30.9	18.1	4.0
Fe				6.6

variable nature of CSH and the points raised previously, there is a good case for assuming that this area consists of calcium silicate hydrates.

- **Area D:** the light grey grain. The analysis of this particle is inconclusive. It could be feldspar but this would have to be confirmed by another technique such as XRD.

Analysis of lightweight concrete

Examining the ×25 magnification images of the lightweight concrete, it is noticeable that the demarcation between aggregate and cement paste is very diffuse. An EDS analysis was carried out on the light granular areas on the backscatter micrograph and these were confirmed as being unhydrated cement particles. The EDS analysis of the centre of the lytag grains has four principal elements: 65.8% O, 11.7% Al, 17.7% Si, 1.3% Ca and 3.4% Fe. This is expressed in terms of the three principal oxides, 51% SiO_2, 28% Al_2O_3, 3.5% CaO and 17% Fe_2O_3. This distribution of oxides is typical of a pozzolan, with high silica and aluminium and low calcium. The distribution of oxides and the presence of iron are indicators confirming that this is a PFA-type material.

Cement–aggregate interface

A good illustration of the nature of the cement–aggregate interface can be seen in Figure 8.14, which is a micrograph of the interface between OPC paste and flint gravel aggregate. A compositional profile, showing atomic percentages of calcium, silica and aluminium, is shown in Figure 8.15.

The profile in Figure 8.15 shows that there is a transition in the silica content across the aggregate boundary. Normally at the aggregate–cement paste boundary there is a higher concentration of Portlandite, that is a low level of silica. In the case of this concrete sample, it is possible that there has been a degree of alkali silica reaction, causing the migration of silica ions from the aggregate into the paste. This sample has been stored in dry conditions for several years and therefore any gel present would have crystallized.

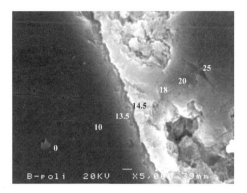

Figure 8.14 *Micrograph (×2500) showing the interface between OPC paste and flint gravel aggregate.*

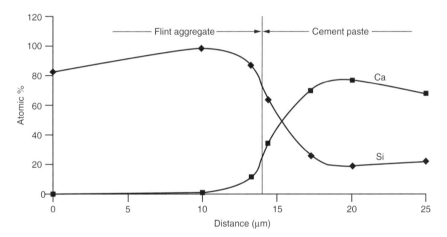

Figure 8.15 *Profile of calcium and silica contents across the aggregate–cement paste interface.*

Water/cement ratio

In this cursory examination by SEM EDS of a sample of concrete, determination of the w/c ratio is not possible; however, one or two conclusions may be drawn. In a review of concrete petrography French[4] suggests two techniques that would indicate the w/c ratio: the distribution of unhydrated cement grains and the SiO_2/CaO ratio. The backscatter image of the samples shows a liberal distribution of unhydrated cement, which would indicate a low w/c ratio. From several EDS analyses of the hydrated paste, the Si/Ca ratio was found to be of the order of 0.3, i.e. $SiO_2/CaO = 0.32$. French's work on standard concrete mixes showed that the lower the SiO_2/CaO ratio, the lower the w/c ratio. From these figures it would appear

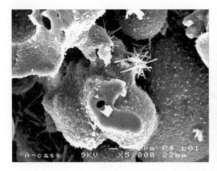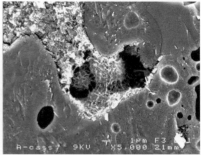

Figure 8.16 *Micrographs of the fracture surface of lightweight concrete. Right: an inclusion in the amorphous sintered PFA (×2500); left: a grain of PFA (×2500).*

that the original w/c content was 0.35–0.4. This low w/c ratio is confirmed by the sparse occurrence of voids and no sign of bleeding.

Morphology of hydrated lightweight concrete

Two micrographs of a fracture surface of the lightweight concrete are shown in Figure 8.16. The images show the structure of the lytag as being amorphous (the glass-like surface in the figure). The voids in the lytag are almost spherical with no interconnections. The growth of fibrous crystals can be seen on the right-hand micrograph.

References

1 ASTM. Test method C1365-98. *Standard method for determination of the proportions of phases in Portland cement and Portland cement clinker using X-ray diffraction analysis*. ASTM, Philadelphia, PA, 2001.
2 Bogue R.H. *The chemistry of Portland cement*. Reinhold, New York, 1947.
3 BSI 6699. *Specification for ground granulated blastfurnace slag for use with Portland cement*. British Standards Institution, 1992.
4 French W.J. Concrete petrography: a review. *Quarterly Journal of Engineering Geology* 1991, **24**, 17–48.

Chapter 9

Case studies

This chapter presents case studies aimed at giving an insight into the methodology for examining deteriorated concrete. The studies have been summarized so that the reader can follow the logic of the investigations without needing to abstract large quantities of data. For reasons of confidentiality the locations of the sites are not given.

Cracking in a reinforced concrete footbridge

Background information

The study concerns a 14 m span reinforced concrete footbridge crossing a river. The bridge is simply supported between two abutments. In cross-section the bridge consists of a deck supported on two webs (Figure 9.1). The bridge is situated in the southern part of the UK and subject to normal external exposure. The water in the river is fresh and there is no possibility of contamination by de-icing salts.

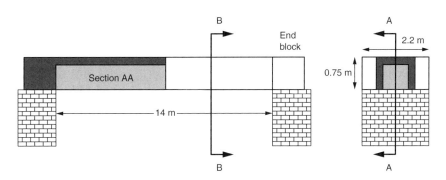

Figure 9.1 *Layout of the bridge (not to scale).*

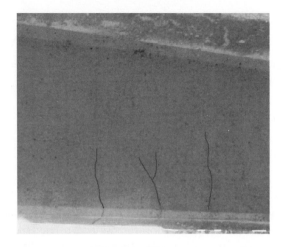

Figure 9.2   *Cracking to the webs of the bridge.*

The problem

An inspection of the structure signalled the presence of vertical cracking in the webs over the span of the bridge. The study aimed to identify the cause of the cracking and verify the quality of the materials.

Inspection

Visual inspection revealed that fine cracks ran vertically from the underside of the webs and that the cracks occurred at regular intervals. Figure 9.2 shows the nature of the cracking and Figure 9.3 illustrates the density of the cracking across the span.

The following inspection strategy was adopted.

- Carry out a covermeter survey to verify the position of the reinforcement.
- Make a break-out to examine the condition of the reinforcement and to conduct an on-site carbonation test.
- Recover dust samples from drilling to determine the chloride profile.
- Take 100 mm diameter cores to enable a petrographic verification of the concrete quality and to examine material in the region of a crack.

Analysis

The covermeter survey indicated that the cover to the vertical link bars was between 25 and 35 mm.

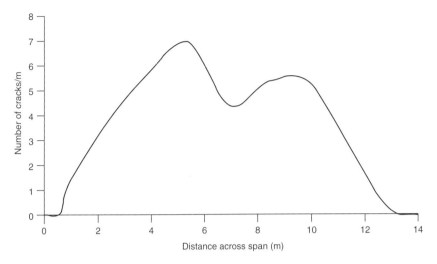

Figure 9.3 *Density of cracking across the span of the bridge.*

A petrographic study of the samples consisted of the following steps.

- The volume proportions of aggregate voids and paste were determined by point counting.
- The water/cement (w/c) ratio was determined petrographically by assessing the unhydrated cement content.

The results indicated that the concrete contained Portland cement at approximately 400 kg/m^3 and a w/c ratio of 0.55. The aggregate consisted of limestone, well graded from 10 mm to dust.

Petrographic examination of the cracks showed that they extended up to about 200 mm from the tension face. The cracks had sharp edges and passed across the sample, traversing both paste and aggregate. There was no sign of deterioration of the concrete along the crack, but it was noted that carbonation had occurred along the line of the cracks up to a depth of about 50 mm.

The break-out of concrete verified that the reinforcement was in good condition and not corroding. A phenolphthalein test in the break-out showed that the depth of carbonation was approximately 3 mm.

The chloride analysis of the dust samples indicated that the chloride content was generally less than 0.1% Cl by weight of cement.

Interpretation

There was no localized deterioration along the cracks and no evidence of corrosion of reinforcement. The cracks passed through both aggregate and cement paste and therefore were not related to plastic cracking. The origin of the cracking appeared to be related to the deflection; the bridge is simply supported and greatest crack density was in the tension zone at midspan. Given the relatively high w/c ratio and the presence of limestone dust, it is also probable that drying shrinkage had occurred. Another possible contributing factor is thermal movement, as the end blocks do not allow for movement.

Having signalled that the concrete is sound, it remains to draw attention to the danger of reinforcement corrosion. Carbonation is progressing along the cracks and will ultimately cause local depassivation. The cover of 25 mm is also somewhat inadequate for the exposure condition.

Recommendations

- **'Do nothing' option:** if no action is taken there is a risk of corrosion to reinforcement occurring at localized sites where the cracks intersect the steel. This will initially lead to rust staining and eventually the integrity of the structure may be threatened.
- **Repair option:** because of the end restraint on the structure a structural appraisal should be carried out. The loss of durability due to reinforcement corrosion can be addressed by crack injection and the application of an elastomeric anti-carbonation coating.

Foundation base cracking

Background information

This study relates to the foundations of a multi-storey structure built over a disused dock. The structure, located in the southern UK, was built in the 1970s. The construction is somewhat unusual: *in situ* concrete foundation bases and columns, together with retaining walls and shear walls were constructed in the bottom of the dock, thus creating a large void beneath the multi-storey structure. These foundation elements were essentially mass concrete and therefore only lightly reinforced. The foundation void was periodically flooded with sea water during the life of the structure, thus causing the build-up of silt. An intended change of use of the void beneath the multi-storey structure necessitated an inspection of the foundation elements.

The problem

During the inspection the silt layer was removed and extensive cracking in the concrete bases was revealed. Because of the level of deterioration and the proposed change of use, the integrity of the bases was brought into question. An investigation was launched to find the cause of the cracking and to make a prognosis concerning the future behaviour of the foundations.

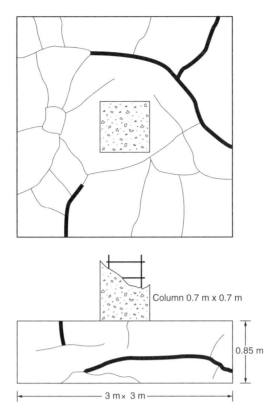

Figure 9.4 *Crack damage to a foundation base.*

Inspection

Site testing carried out on the bases consisted of an initial dimensional survey to record the crack damage and structural details. The cracking pattern found on the bases is illustrated in Figure 9.4. To evaluate the condition of the reinforcement a half-cell potential survey was carried out on the bases together with break-outs of material to examine the reinforcement *in situ*. Dust samples obtained by drilling were used to determine the chloride ingress profile in the concrete. Core samples were taken from the bases to be used for the determination of mix characteristics, concrete quality and cause of the cracking.

Analysis

The site testing indicated that the chloride concentration at the surface was quite high, up to 1.3% Cl by weight of cement; however, this concentration fell off rapidly with depth and critical levels were not attained at the level of the steel. The half-cell testing indicated areas of high potential, which were attributed to the presence of dampness in the concrete. None of the break-outs showed signs of corrosion.

Petrographic analysis and strength testing of the cores indicated a good quality concrete:

- concrete strength (expressed as *in situ* cube strength): 40–57 N/mm^2
- w/c ratio: 0.44–0.52
- average cement content: 450 kg/m^3
- a dense, robust, re-crystallized limestone aggregate showing no sign of deterioration
- the cement used was ordinary Portland (OPC); no trace of pulverized fuel ash (PFA) or ground-granulated blast furnace slag (GGBFS) was found.

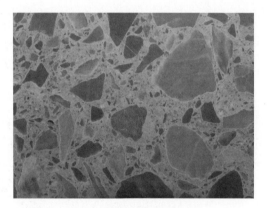

Figure 9.5 *General appearance of the concrete: the cracks, visible with a hand lens, are found around the periphery of the aggregate particles.*

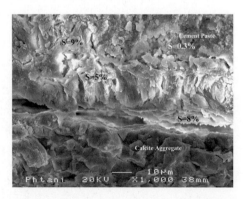

Figure 9.6 *Micrograph showing the area of the fissure between the cement paste and the aggregate. The sulfate concentrations marked on the micrograph show the high concentration of sulfur in the zone of the crack. A needle-like curtain of crystals, probably ettringite, can be seen along the upper edge of the fissure.*

The macroscopic cracks at the surface were lined with deposits of calcite and magnesium compounds, most probably resulting from the contact with sea water. The general appearance of the concrete is shown in Figure 9.5. A micrograph of the fissure between the aggregate and the cement paste is shown in Figure 9.6. Energy dispersive X-ray spectrometry (EDS) was carried out on the paste and the crystalline forms seen in the zone of the fissure.

Examination of these cracks also revealed the presence of needle-like crystal formations. Thin-section petrographic analysis showed ettringite around the periphery of aggregate particles, and in cracks and partings in the cement paste. To confirm the presence of ettringite, samples were examined using scanning electron microscopy (SEM) and EDS. Table 9.1 sets out the average compositional data for the concrete; the EDS system was calibrated using mineral and cement standards. Table 9.2 shows the composition of the crystal forms filling the voids; examination of the observed and theoretical oxide compositions confirms that the needle-like forms are ettringite.

Several samples were tested for expansion using the procedure set out by the BCA.[1] Although this test is designed to detect the alkali silica reaction (ASR), which was not seen in any cores, continued expansion due to delayed ettringite formation (DEF) is possible. The expansions measured ranged from zero to 6 mm/m.

Table 9.1 *Compositional analysis of the concrete (normalized weight percentage of oxides)*

Oxide	% by mass
SiO_2	19.65
TiO_2	0.41
Al_2O_3	5.48
Fe_2O_3	2.67
Mn_2O_3	0.03
MgO	1.34
CaO	66.66
Na_2O	0.37
K_2O	1.35
SO_3	1.98
Cl	0.05
P_2O_5	0.09

Total alkali (Na_2Oe) = K_2O + 0.658 Na_2O.

Table 9.2 *EDS analysis of material filling cracks (normalized weight percentage of oxides)*

Oxide	% by mass	Ettringite theoretical
SiO_2	0.89	
TiO_2	0.05	
Al_2O_3	13.35	15
Fe_2O_3	0.15	
Mn_2O_3	0.02	
MgO	0.12	
CaO	56.41	50
Na_2O	0.11	
K_2O	0.27	
SO_3	29.01	35
Cl	0.06	
P_2O_5	0.19	

Table 9.3 *Summary of oxides found in cracked and non-cracked samples*

Oxide	Cracks	No cracks
MgO	1.22	1.12
Al_2O_3	5.65	5.80
Na_2O	0.35	0.32
K_2O	1.54	0.94
SO_3	1.67	2.23
Total alkali (Na_2O eq)	1.78	1.15
SO_3/Al_2O_3	0.30	0.39

Interpretation

The concentration of SO_3 found in the concrete was normal and therefore ettringite formation due to an external sulfate source can be ruled out. It is evident from the nature of the formations of ettringite that the deterioration of the concrete is caused by DEF and secondary ettringite formation where sea water had penetrated thermal cracks in the concrete. DEF is normally associated with steam-cured concretes, ettringite being unstable at temperatures exceeding 60–70°C. At the time of construction the ambient temperature was exceptionally high, up to 30°C, so it is possible with the heat of hydration in the mass concrete bases that the temperature in the core of the base could have achieved values associated with steam curing. The risk of DEF cracking occurring has also been linked to the following factors:[2] sulfate content, magnesium content, total alkali content and cement fineness. The ratio SO_3/Al_2O_3 has also been cited as a factor; for cements with $SO_3/Al_2O_3 < 0.55$ the risk of DEF is reduced.[3] Table 9.3 reports an analysis of the oxide concentration cited as involved in DEF; the table groups the samples into those where cracking was found and those without cracks. It can be seen from this cursory analysis that there is a trend confirming conditions conducive to DEF, i.e. in the cracked samples the magnesium content and the alkalinity were higher than the non-cracked samples. This difference in alkalinity may suggest that two cement sources have been used. The sulfate/alumina ratio was lower than the suggested limit.

The observed crack damage to the bases is a complex phenomenon and probably linked to three main factors:

- thermal cracking resulting from the high temperature due to the high cement content, plus elevated ambient temperature at the time of construction
- DEF: the ettringite formations seen in the petrographic study are typical of DEF. The samples suffering from cracking had high alkalinity, and high alkalinity cements are often associated with DEF
- secondary ettringite formation in the cracks penetrated by sea water.

Recommendations

In view of the extent of internal cracking to the base slabs a reappraisal of the structural integrity of the building is essential. Two possible options could be considered after the appraisal:

- downgrading of the loading of the structure
- re-designing, upgrading and replacement of the deteriorated foundation pads.

It is essential, whatever remedial steps are taken, to ensure that the concrete remains dry to prevent further expansion.

Marine structure

Background information

The structure consists of a reinforced concrete quay constructed in the mid-1970s, located on the north French coast. The structure consists of a reinforced concrete deck slab, supported on columns, 0.6 m × 0.6 m in section and 4 m high. Half-cell potential testing had been carried out on the soffit and column heads in the 1990s. The structure is exposed to marine conditions: the tidal limit in this location is approximately 6 m.

The problem

The inspection was carried out because of concern about the level of damage resulting from apparent corrosion to the reinforcement. This had progressed as far as cracking to columns, and there was also spalling and loss of cover to the soffit of the deck.

Figure 9.7 *Detail of the damage to a column. Note the heavy rust staining and cracking.*

Inspection

The testing and sampling were carried out on appropriately sized test areas, which were marked out on a 0.5 m grid for reference. Half-cell potential and cover to steel were measured at each of the grid nodes. Dust samples were collected by drilling to obtain the chloride concentration profile and depth of carbonation was determined from freshly fractured samples. Cores were taken from representative parts of the structure.

Visual inspection revealed cracking and heavy rust staining on the columns (Figure 9.7). Similarly, the soffits of the slabs exhibited signs of corrosion damage, spalling and, in places, a total loss of the cover concrete (Figure 9.8).

Analysis

Petrographic analysis of the cores indicated that the concrete in the structure was generally of good quality.

- The concrete in the structure was based on OPC with no trace of replacements.

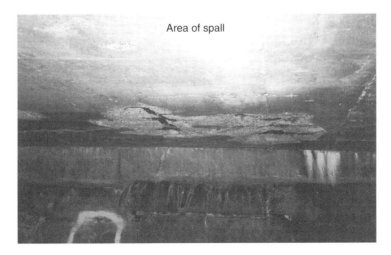

Figure 9.8 *Detail showing the loss of cover on the underside of the deck.*

- The cement content ranged between 310 and 360 kg/m^3, the w/c ratio was 0.54–0.56 and concrete strength (expressed as *in situ* cube strength) was 52–69 MPa.
- The coarse aggregate consisted of a basic igneous rock, which showed no sign of deterioration.

Some evidence of ASR was found but the incidence was restricted to certain sections of the structure. Petrographic analysis of the cores showed small quantities of ettringite in the voids of the cement paste. This was taken as evidence of movement of minor quantities of water through the concrete.

The covermeter survey was carried out using an electromagnetic covermeter, following the method set out in BS 1881: Part 204: 1988. The covermeter used had simultaneous correction for bar size and lapped or twinned bars. The minimum cover to steel was 32 mm, with an average cover of 55 mm.

The depths of carbonation were determined on-site by spraying freshly fractured surfaces with phenolphthalein solution. The depths of carbonation varied from 0 to 10 mm, but most of the depths were less than 5 mm.

The half-cell potential survey was carried out in accordance with the ASTM procedure.[4] A silver/silver chloride (Ag/AgCl) half cell was used, which gives potential readings approximately 50 mV more positive than the copper/copper sulfate half cell. Figure 9.9 shows the correspondence between the spalled and low-cover areas and zones of high potential.

The dust samples for chloride analysis were collected in accordance with BS 1881: Part 124: 1988. The sampling depths for the dust were 5–30 mm, 30–55 mm, 55–105 mm and 105–130 mm. The chloride content was determined by laboratory testing. Figure 9.10 shows a typical chloride profile for the quay. Assuming an average cover of 55 mm, a statistical

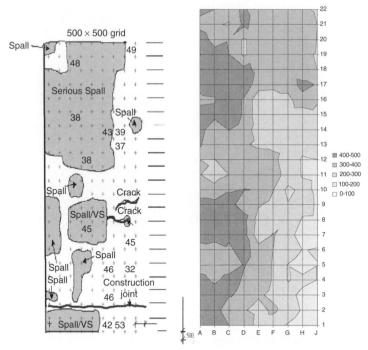

Figure 9.9 *Results of the covermeter survey (left) and the half-cell (Ag/AgCl) potential survey (right). The grid interval used is 0.5 m. Note: if the potential is more negative than −350 mV Cu/CuSO₄ (−300 mV Ag/AgCl) the risk of corrosion is designated as high.*

analysis of the chloride profiles showed that 60% of the surveyed sample had a chloride content (chloride % of cement) greater than 0.4%. The chloride content in the columns was greater than that found in the soffit of the deck slab.

Interpretation

The petrographic analysis indicated that the mix used had a cement content of about $350 \, \text{kg/m}^3$ and a w/c ratio of 0.55. The mean cover to steel on the structure was 55 mm, with a minimum of 32 mm. BS 8110: 1985 would classify the exposure environment as 'extreme': Table 3.4 in BS 8110 would require 60 mm cover for a C 45 grade concrete of $350 \, \text{kg/m}^3$ and a w/c ratio of 0.5. The exposure condition is classed as most severe by BS 8110: Part 1: 1997; this would call for a nominal cover of 50 mm, minimum of grade C50 concrete, cement content of $400 \, \text{kg/m}^3$ and w/c ratio of 0.45.

The concrete found on the structure was sound, with the exception of one area. The chloride profiles indicated that a significant proportion

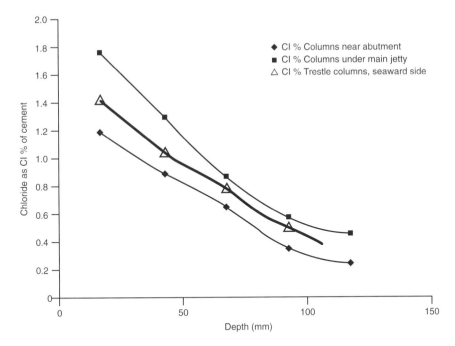

Figure 9.10 *Typical chloride profile for the structure.*

of the reinforcement was exposed to chloride concentration liable to cause depassivation. Chloride ingress and consequential corrosion of the reinforcement was seen as the principal cause of deterioration to the structure.

Recommendations

- **'Do nothing' option:** given the levels of damage apparent and the high levels of chloride in the concrete, the extent of corrosion damage will intensify rapidly. This will lead to severe loss of section to the columns and areas of the soffit.
- **Patch repair option:** patch repairs could not be used owing to the likelihood of incipient anode formation. This would result in new corrosion cells appearing around the patch repaired area. Early failure would result.
- **Repair and cathodic protection option:** the damaged areas of the structure could be repaired and a cathodic protection system installed. In areas of the structure where chloride levels are low, i.e. some areas of the soffit, a corrosion monitoring system should be installed.

Deterioration of foul sewers

Background information

An urban agglomeration in the southern UK consists of residential and industrial developments. Wastewater from the area is transferred to a sewage treatment works by a 10 km trunk gravity sewer with sections of rising mains within the catchment. The trunk sewer was constructed in the 1930s; however, various sections have been replaced on a piecemeal basis. The diameter of the sewer varies from 600 to 1200 mm; the smaller diameter pipes are in salt glazeware and the larger diameters in concrete. The 5 km section of sewer upstream of the sewage works was constructed in the late 1950s and consists of 1200 mm spun concrete pipework with manholes constructed in concrete. The wastewater has a high industrial content, notably wastes from detergent and food processing industries. This type of wastewater has a high organic and sulfate content.

The problem

Since the 1980s routine sewer inspection has highlighted internal deterioration of the pipes and concrete manholes. The sewer system has, for many years, been subject to sediment build-up in areas of slack gradient and also prone to the build-up of high concentrations of hydrogen sulfide gas. Lack of capacity due to growth and several sewer collapses triggered the need for an assessment of the condition and possible replacement of the sewer.

Inspection

Visual inspections of manholes revealed that the loss of concrete section had progressed as far as the reinforcement. There was no surface deposit covering the deteriorated concrete. Particularly severe concrete deterioration was seen where *in situ* concrete had been used. In these zones the concrete had no superficial loss of section, but it had deteriorated internally to the point where it could be removed by hand. The exposed concrete was highly fissured and the cement paste had totally degenerated.

Damage to the concrete pipework was severe in places. In one tunnelled section of the sewer, the concrete had disintegrated to expose the chalk. In all cases the most severe damage was found above the normal water level, whereas damage to the invert was minimal or non-existent.

Analysis

Two samples were recovered for analysis: one sample from the pipe section and the other from *in situ* placed concrete from below the cover of a manhole.

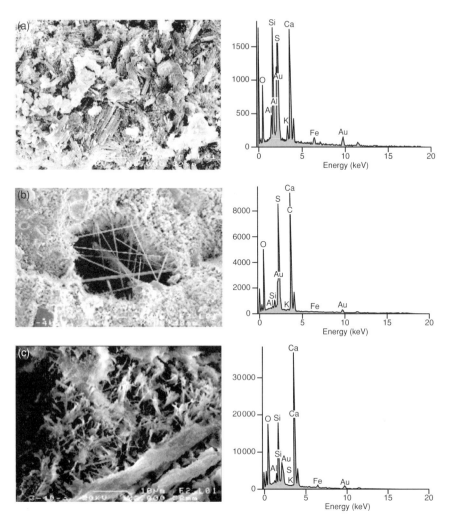

Figure 9.11 *SEM and EDS spectra for concrete attacked by sulfuric acid. (a) Superficial layer of gypsum (×1000); (b) voids filled with gypsum at 1 mm from the surface (×250); (c) normal hydrated cement at 2 mm from the surface (×1000). It can be seen from the three sections that the formation of gypsum is restricted to a depth of less than 2 mm.*

A section of failed concrete pipe was analysed to determine the extent and nature of the damage. Visual inspection of the pipe showed that the exterior was in almost perfect condition, whereas the internal section had been significantly reduced at the 10 o'clock and 2 o'clock positions. The internal surface of the pipe had a white–grey deposit to a depth of about 1 mm; behind this layer the concrete had a homogeneous appearance.

Powder samples were taken from the internal surface of the pipe and from concrete behind the deteriorated layer. X-ray diffraction (XRD) analysis showed that the deteriorated zone consisted solely of gypsum and quartz. Behind the deteriorated layer, XRD analysis indicated that the concrete was sound; no abnormal quantity of gypsum or ettringite was found. Figure 9.11 shows diffractograms and SEM images of these two zones.

Interpretation

Under anaerobic conditions, occurring in rising mains or in semi-stagnant flow conditions, sulfates and sulfites will be reduced to hydrogen sulfide. When this wastewater enters a turbulent section of the sewer hydrogen sulfide gas will be liberated. Within the sewer this gas was oxidized by aerobic *Thiobacillus* bacteria to sulfuric acid (pH ~ 2). The sulfuric acid formed behind the bacterial layer attacked principally the Portlandite, thereby producing gypsum.[5] The conditions within the sewer, aerated, with a high H_2S gas concentration and high humidity, were conducive to the progression of this reaction.

The situation in the manholes, notably in the upper parts, was somewhat different. Here the relative humidity was much lower and there was little or no evidence of a bacterial slime layer. XRD analysis of the deteriorated *in situ* concrete indicated that there had been a conversion of the Portlandite to gypsum. The mechanism of deterioration is possibly due to direct reaction between H_2S gas and the Portlandite. The high porosity of the *in situ* concrete allows oxygen and H_2S gas migration into the pore structure. The formation of gypsum can then occur:

$$H_2S + Ca(OH)_2 + 2O_2 \rightarrow CaSO_4 \cdot 2H_2O$$

The gypsum is expansive and will cause internal cracking in the concrete.

Recommendations

- **'Do nothing' option:** This would be impractical as the sewer is rated as 'critical' and the risk of collapse is high.
- **Rehabilitation option:**[6] Where capacity is adequate internal lining can be installed. Because of the poor structural condition of the existing pipe these liners would have to be load bearing. In man-entry sewers, grouted-in glass-reinforced plastic liners could be used. In the case of small-diameter sewers several relining techniques exist. In the case of rising mains, polyethylene pipe can be pulled through the existing pipe, the reduction in roughness often compensating for the reduction in cross-sectional area.

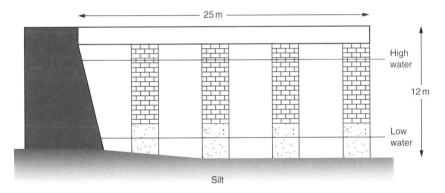

Figure 9.12 *Layout of the wharf structure.*

- **Replacement option:** Where the hydraulic capacity of the sewer system is in question, the best solution is sewer replacement, preferably with a lined concrete pipe (e.g. epoxy coated). In a developed area tunnelling is often a preferred option owing to problems with access and service diversions.

Dock structure

Background information

The structure is a pier that was added to an existing dock wall to extend the frontage of the berth. The structure is located in an estuary to the west of the English Channel, the water is tidal and the salinity very close to that of normal sea water. The original structure dates from the mid nineteenth century and is constructed in granite blockwork. The extension to the wharf was added in the 1930s and consists of four abutments supporting a reinforced concrete deck. Key dimensions and layout are shown in Figure 9.12. The abutments were built in two stages, *in situ* concrete foundations and abutments constructed in mass concrete 0.5 m cubic blocks.

The problem

Routine inspection had previously highlighted extensive chloride damage to the reinforced concrete decking. The abutments above water level were noted to be in good condition, the concrete was sound but there was some opening of the joints. The concrete at low water level was deteriorated, therefore, before making a decision about the repair strategy, it was decided to undertake an underwater survey to verify the condition of the lower part of the abutments.

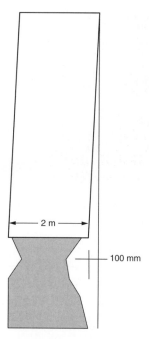

Figure 9.13 *Vertical section of an abutment.*

Inspection

The extent of the damage to the abutments became apparent on the first reconnaissance dive, during which mass erosion of the concrete beneath the blockwork was found. The survey was carried out using tape surveys to determine the vertical profile of the abutments. This was followed by removal of samples for analysis, CCTV survey and still photographs. The diving conditions in the port were very poor owing to high silt loads caused by tide movement and shipping movements, and visibility was often zero. High water and slack tide afforded the best conditions for a survey; visibility under these conditions could be up to 1 m. A typical profile of the end abutment is shown in Figure 9.13. The erosion of the *in situ* foundation has led to an undercutting blockwork and the face of the seaward abutment is out of vertical. The concrete beneath the blockwork was heavily deteriorated, and in most places it could be removed with hand tools. In places the deteriorated face of the abutment was exfoliating in large sections. Samples of the deteriorated concrete were removed for analysis.

Interpretation

The recovered samples of the lower abutments showed that the cement paste had deteriorated to a white, paste-like material. A relatively sound

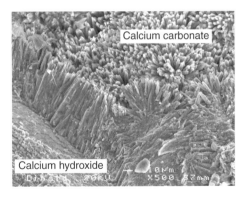

Figure 9.14 *SEM of the deteriorated concrete (×200).*

sample was cored and its strength assessed by a point load test,[7] which indicated an effective cube strength of 25 N/mm².

A more detailed visual inspection of the samples showed that the aggregate was a mixture of quartz and crushed granite. There was a large fraction of shells, which indicated that beach sand, unwashed and ungraded, had been used in the mix. A sample of the deteriorated material was examined by XRD and SEM (Figure 9.14). Given the obvious condition of the concrete this analysis was not necessary; however, it is reported to illustrate the mechanisms involved in sea-water attack. The XRD analysis of the concrete attacked by sea water showed several crystal species co-existing in the sample, the most common being quartz. Strong concentrations of calcium carbonate, possibly aragonite, were found, as well as traces of ettringite and thaumasite.

Concrete deterioration in the presence of sea water is complex. The deteriorated zone is just above and below low-tide level, where several vectors of attack can occur:

- erosion due to water movement
- freeze–thaw action
- magnesium ion exchange
- sulfate attack
- carbonation.

In this case it is evident that there has been a high degree of conversion of the lime to carbonate. In sea water, aragonite, orthorhombic $CaCO_3$, forms in preference to calcite owing to the presence of magnesium. The high degree of carbonation may result from the high free lime content in the cement used in construction. The aggregate used appears to have been local beach sand. XRD analysis showed the presence of species normally associated with granite and altered species such as kaolin were also identified. Geologically, this area of the French coast consists of weathered granite and highly metamorphosed slates and shales.

The problem of scour to the weakened areas of the abutments was exacerbated by turbulence from ship propellers, which is unavoidable in this kind of installation.

The end abutment appeared to have undergone some rotation, but there was no sign of associated cracking at the deck abutment joint. The deck appeared to be tying the structure back to the old wharf. It would appear, from the survey, that the most likely mode of failure would be a settlement of the end pier due to extensive wash-out of the old foundation. The massive over-design of the abutments has compensated for the very inadequate materials used at the time of construction.

Recommendations

- **'Do nothing' option:** this is unacceptable given the extent of damage. The immediate action must be a weight restriction on the pier structure.
- **Repair option:** given the depth of deterioration to the concrete it would be virtually impossible to envisage a refilling of the voids resulting from the undercutting. Provision of a sheet pile curtain around the structure would stabilize the situation.
- **Replacement option:** there is doubt about the longevity of the extension owing to the extensive chloride damage to the deck. An evaluation of the need for the structure should be undertaken before considering total replacement.

References

1 British Cement Association. Diagnosis of alkali-silica reaction, Appendix H. *Report of a BCA working party*, 2nd edn, British Cement Association, London, 1992.
2 Hobbs D.W. Ettringite: the sometimes host of destruction. *ACI Special Publication SP177*, 1999.
3 Odler I. and Chen Y. On the delayed expansion of heat cured Portland cement pastes and concretes. *Cement and Concrete Composites* 1996, **18**, 181–185.
4 ASTM C876-80. *Standard test method for half cell potentials of reinforcing steel in concrete*. American Society for Testing and Materials, Philadelphia, PA, 1980.
5 Thistlethwayte D.K.B. *The control of sulphides in sewerage systems*. Butterworths, London, 1972.
6 *Sewerage Rehabilitation Manual Vols 1 and 2*, 3rd edn. WRc plc, Medmenham, UK, 1994.
7 ASTM D5731-95. *Standard test method for determination of the point load strength index of rock*. American Society for Testing and Materials, Philadelphia, PA, 1995.

Appendix 1

Data interpretation

Indicators of concrete integrity

	Units	Good	Average	Poor
Porosity		$<10\%$	$10–15\%$	$>15\%$
Water permeability	m/s	$<10^{-12}$	$10^{-12}–10^{-10}$	$>10^{-10}$
Gas permeability	m/s	$<2 \times 10^{-18}$	2×10^{-18} to 2×10^{-17}	$>2 \times 10^{-17}$
Gas diffusion coefficient	m²/s	$<5 \times 10^{-8}$	5×10^{-8} to 50×10^{-8}	$>50 \times 10^{-8}$
Chloride ion diffusion	m²/s	$<1 \times 10^{-12}$	1×10^{-12} to 5×10^{-12}	$>5 \times 10^{-12}$
ISAT at 30 min	ml/m² · s	<0.17	$0.17–0.35$	>0.35
ISAT at 1 h	ml/m² · s	<0.10	$0.10–0.20$	>0.20

Based on data from Concrete Society Technical Report 54.

Indicators of reinforcement corrosion

Resistivity

Resistivity readings are dependent on the moisture content of the concrete and therefore values may change from day to day. For guidance, the following interpretations of resistivity measurements have been cited when referring to depassivated steel.

Resistivity (Ωm)	Corrosion rate
>200	Low
$100–200$	Low to moderate
$50–100$	High
<50	Very high

Chloride content

Chloride content by weight of cement (%)	Condition of concrete	Risk of corrosion
>1.0	All concrete	High
0.4–1.0	Carbonated concrete	High
	Uncarbonated with $C_3A < 8\%$	High
	Uncarbonated with $C_3A > 8\%$	Moderate
<0.4	Carbonated concrete	High
	Uncarbonated with $C_3A < 8\%$	Moderate
	Uncarbonated with $C_3A > 8\%$	Low

Half-cell potential

Half-cell potential (mV) $Cu/CuSO_4$ half cell	Risk of corrosion
>−200	Low
−200 to −350	Moderate
>−350	High

Note that the half-cell potential only indicates what is happening on the day of measurement. Measurements during dry summer periods can give misleading results. Actual corrosion damage also depends heavily on the resistivity of the concrete.

Linear polarization

The following broad criteria for corrosion have been developed from field and laboratory investigations with the sensor-controlled guard ring device.

I_{corr} ($\mu A/cm^2$)	Corrosion
<0.1	Passive condition
0.1–0.5	Low to moderate corrosion
0.5–1.0	Moderate to high corrosion

Indicators of concrete strength and quality

Ultrasound

The relationship between the velocity of the ultrasound pulse and the concrete strength is highly dependent on the mix used. Figure A1.1 gives an indication of cube strength as a function of ultrasound pulse velocity (UPV).

Schmidt hammer

Figure A1.2 shows an example of the relationship between rebound hammer results and the cube strength of the concrete. It should be remembered

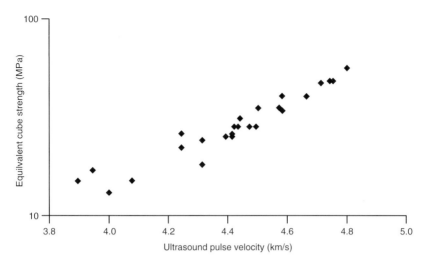

Figure A1.1 *Typical relationship between cube strength and ultrasound pulse velocity (UPV).*

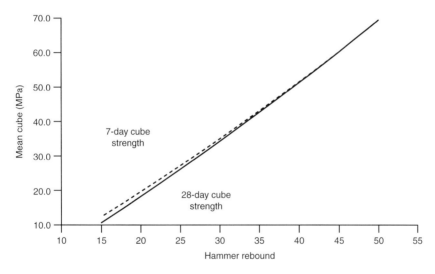

Figure A1.2 *Relationship between rebound hammer reading and the cube strength of a concrete. The compressive strength for a cylinder = 0.85 × cube strength.*

that this type of relationship is very much dependent on the concrete constituents and mix, i.e. the calibration will vary from one concrete to another. The Schmidt hammer, like UPV testing, is very good at comparing two areas of concrete but attempts to measure strength indirectly require careful calibration.

Fire-damaged concrete

Changes in fire-damaged concrete

Temperature (°C)	Changes
<300	Boundary cracking alone
250–300	Aggregate colour changes from pink to red
300	Paste develops a brown or pinkish colour
300–500	Serious cracking in paste
400–450	Portlandite converts to lime
500	Change to anisotropic paste
500–600	Paste changes from red or purple to grey
573	Quartz gives a rapid expansion resulting from a phase change from alpha to beta quartz
600–750	Limestone particles become chalky white
900	Carbonates start to shrink
950–1000	Paste changes from grey to buff

Changes in aggregate

Temperature (°C)	Changes
250–300	Aggregate colour changes from pink to red
573	Quartz gives a rapid expansion resulting from a phase change from alpha to beta quartz
600–750	Limestone particles become chalky white
900	Carbonates start to shrink

Changes in the paste

Temperature (°C)	Changes
300	Paste develops a brown or pinkish colour
400–450	Portlandite converts to lime
500–600	Paste changes from red or purple to grey
950–1000	Paste changes from grey to buff

Cracking

Temperature (°C)	Changes
<300	Boundary cracking alone
300–500	Serious cracking in paste
500	Change to anisotropic paste

Appendix 2

Interaction between radiation and a solid

This appendix explains, in simple terms, some of the basic concepts encountered in scanning electron microscopy, micro-analysis and X-ray diffraction (XRD) analysis. The discussion is centred on the interaction between a beam of radiation and a solid; within the context of this book the interaction of X-rays and electrons with a solid will be discussed in some detail.

Basic principles

Relationship between energy and wavelength

Definitions

E energy of the photon: expressed in electron volts (eV) or joules (J)
f frequency: expressed in nanometres (nm) or Ångström units (Å)
h Planck's constant: 6.62×10^{-34} J·s or 4.14×10^{-15} eV·s
c velocity of light: 3×10^8 m/s
e electron charge: 1.6021×10^{-19} C

Electromagnetic radiation
There is an important relationship between the energy, wavelength and frequency of electromagnetic energy:

$$E = hf = \frac{hc}{\lambda} \tag{A2.1}$$

X-rays used in the study of crystal structure are in the range of:
Wavelength   0.2–2 Å
Energy   60–6 keV

Electrons
An electron is a particle which at rest can be attributed with a mass m_o and an electron charge e. If the electron is accelerated by a voltage difference V it has a kinetic energy of $E = \frac{1}{2}m_o v^2 = eV$, v being the velocity.

Maurice de Broglie showed that moving electrons have wave properties and their wavelength can be determined from the particle momentum:

$$\lambda = \frac{h}{mv} = \frac{h}{\sqrt{2mE}}$$

correcting the mass for relativity,

$$m = \frac{m_o}{1 - \left(\dfrac{v^2}{c^2}\right)}$$

and substituting for the universal constants and expressing E in electron volts:

$$\lambda = \frac{12.26}{\sqrt{E(1 + 0.979 \times 10^{-6}E)}}$$

where E is energy (eV) and λ is the wavelength (Å).

For example, a typical scanning electron microscope has a 30 keV electron beam. Applying the above equation the wavelength of the beam is found to be 0.07 Å.

Atomic structure

In the following sections reference will be made to energy levels of electrons in an atom. The electron structure around the nucleus of an atom can be expressed in terms of four quantum numbers:

- the principal quantum number (n) describing the distance of the electron shell from the nucleus: $n = 1$ (K shell), 2 (L shell), 3 (M shell), etc.
- the Azimuthal quantum number (l), which describes the shape of the orbit, represented by the letters s, p, d and f: $l = 0, 1, 2, ..., n - 1$
- the magnetic quantum number (m), which describes the orientation of the orbit relative to the nucleus: $-l < m < +l$
- the spin quantum number (s), which can be envisaged as the spin of the electron about its own axis: $s = \pm \frac{1}{2}$

Each element has a unique arrangement of electrons, which can be described in terms of the quantum numbers.

The notion of energy levels is important when discussing the interaction between radiation and a solid. To determine the number of energy states in an atom it is convenient to introduce a new quantum number j which corresponds to the total angular momentum of the electron: $j = l + s$

Shell	n	m	Sub-layer	l	$j = l \pm 1/2$	Number of electron states $2j+1$	Energy level	
K	1	0	s	0	1/2	2	K	—
L	2	0	s	0	1/2	2	L1	—
		−1,0,+1	p	1	1/2	2	L2	—
					3/2	4	L3	—
M	3	0	s	0		2	M1	—
		−1,0,+1	p	1	1/2	2	M2	—
					3/2	4	M3	—
		−2,−1,0,1,2	d	2	3/2	4	M4	—
					5/2	6	M5	—

(Energy increasing)

Figure A2.1   *Electronic configuration and energy levels for an atom.*

$(j \neq -\frac{1}{2})$. The number of electron states for each value of l is given by $(2j + 1)$ (Figure A2.1).

Electron–solid interactions

When a beam of radiation passes through a material, the radiation may undergo a modification as will, under certain conditions, the material. The incident ray may undergo a diminution of intensity, loss of energy (absorption) and a change of direction (scattering). The modification to the material will result in a transfer of energy from the radiation to the atoms in the material. There are two types of interaction:

- **elastic interactions:** where the internal structure and energy of the atom is unchanged or slightly changed. Therefore, the energy and the wavelength of the beam are unchanged in the interaction; however, the radiation may be scattered
- **inelastic interactions:** where the atoms in the structure undergo an energy change and consequently the energy of the radiation is diminished in the interaction.

Elastic interactions

The following elastic interactions are possible when a beam of electrons bombards a material:

- **thermal effects:** a small transfer of energy to the atom causes vibration, which is manifested as a production of heat
- **chemical effects:** breaking of chemical bonds within the material
- **atomic displacement:** if the transfer of energy from the electron to the atom is greater than the binding energy between two atoms it is possible to displace atoms permanently, i.e. cause radiation damage

- **scattering:** at certain angles of incidence a beam of electrons will be elastically deflected. This is observed as the phenomenon of backscattering.

Inelastic interactions

Deceleration
Deceleration of an electron occurs as it passes through and interacts with the other electrons in the atomic structure. This results in a loss of kinetic energy, which is translated into emissions of X-rays. Because of the random nature of these interactions the emitted X-rays have a wide spectrum of wavelengths.

Excitation of the atom at a core level
This form of inelastic interaction is fundamental to many of the techniques used in micro-analysis and therefore is explained at greater length. When an electron beam hits a material, most of the energy is dissipated thermally; however, about 1 in 10000 electrons will manage to displace one of the core electrons at level K or L. This excitation of the atom due to the displaced electron is relaxed by one of the following mechanisms.

- There is an ionization of the atom due to the emission of a core level electron, at the K level, termed **secondary electron** emission (Figure A2.2a).
- To maintain equilibrium there is a de-energization (relaxation) of the atom. One of the outer shell electrons, $L3$ for example, replaces the displaced electron. The outer electron must fill the internal shell, the difference in energy level corresponds to the energy of the emitted X-ray (**X-ray fluorescence**). As each element has a unique atomic structure it follows that the energy level of the X-ray emission is characteristic of the element (Figure A2.2b).
- The relaxation of an excited atom can also occur by the displacement of an electron from a core level, $L1$, to replace the electron lost from the K level. To maintain the balance an electron in the outer layer is released. This phenomenon is known as **Auger electron emission**, an effect observed by Pierre V. Auger in the 1920s. The energy levels of Auger electrons are characteristic of the elements present and the quantity of Auger electrons is proportional to the concentration of atoms at the surface (Figure A2.2c).

X-ray emission or Auger electron emission?

In the preceding section it was seen that the excitation of an atom by the removal of a core electron can result in the emission of either a photon of X-ray or an Auger electron. The energy of both of these emissions is

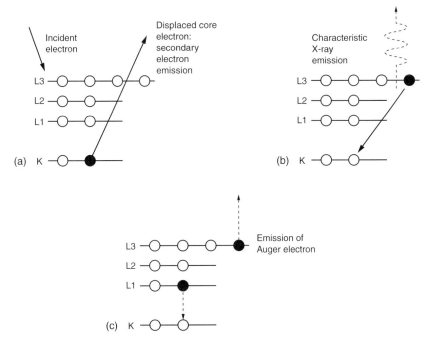

Figure A2.2 *(a) Ionization of the atom and release of secondary electrons. (b) Core level electron replacement with the emission of X-rays. (c) In the ionized atom, a transition from L1 to K occurs to replace the core electron and the Auger electron is emitted.*

characteristic of an atomic structure, i.e. unique to an element; therefore, the measurement of the energy of the emission is a potential means of determining the chemical composition of a material. The energy of these emissions is linked to the atomic number of the element undergoing the interaction: the form of the emission, an Auger electron or a photon of X-ray, must be determined. Figure A2.3 shows that the probability of the emission of an Auger electron is higher for elements with a low atomic number, whereas the converse is true for X-ray emission.

X-ray fluorescence

The emission of X-rays due to inelastic interactions is fundamental to X-ray fluorescence (XRF), X-ray diffraction (in generating the beam of the X-ray) and energy dispersive X-ray spectrometry (EDS). When a beam of high-energy electrons bombards a target, a series of interactions may take place between the electron and the atoms in the target. One possible effect is the production of X-rays (see Figure A2.2b). As an electron enters a solid it will undergo a deceleration due to a series of inelastic interactions with atoms in

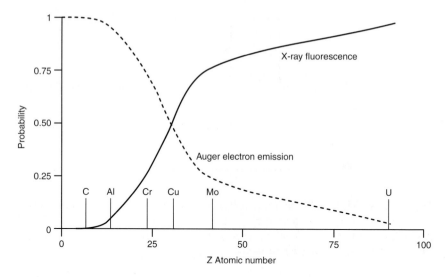

Figure A2.3 *Probability of a specific emission with respect to the atomic number.*

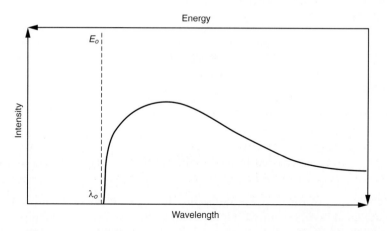

Figure A2.4 *X-ray spectrum due to the deceleration of an electron within a solid.*

the material. With each interaction the electron loses energy which is converted into an emission of electromagnetic radiation, in this case X-ray
emissions. The wavelength of the X-rays emitted with each interaction is
directly related to the energy change in the electron. Owing to the incre-
mental energy loss during deceleration there is an emission of a continuous
spectrum of X-rays. This continuous spectrum is illustrated in Figure A2.4.
The shortest wavelength of radiation that can be emitted will be governed
by the energy of the incident electron beam E_o; if the electron were to

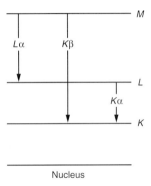

Nucleus

Figure A2.5   M, L *and* K *are the first three electron shells. If an electron is displaced from* K *it can be replaced by an electron in either* L *or* M. *A different energy level is associated with each shell; therefore, the frequency of the X-ray emission will be fixed by the change in level, i.e.* Kα, Kβ *or* Lα.

undergo total deceleration with its first interaction with an atom it would undergo an abrupt energy change of E_o producing radiation at a wavelength of λ_o:

$$\lambda_o = \frac{hc}{E_o}$$

The second method by which the high-energy incident electron can interact with an atom is by displacing one of the inner level electrons. To restore a stable state in the atom one of the outer shell electrons moves to replace the displaced electron (Figure A2.2b). The wavelength of the X-ray emitted can be calculated from the difference in energy between the two electron levels. Figure A2.5 illustrates a simplified version of the possible electron transitions. Knowing this difference in energy, the wavelength of the emitted X-rays can be calculated:

$$E_2 - E_1 = \frac{hc}{\lambda}$$

	ΔE (J)		
$L \rightarrow K$	1.29×10^{-15}	Kα	$1.54\,\text{Å}$
$M \rightarrow K$	1.43×10^{-15}	Kβ	$1.39\,\text{Å}$
$M \rightarrow L$	0.15×10^{-15}	Lα	$13.36\,\text{Å}$

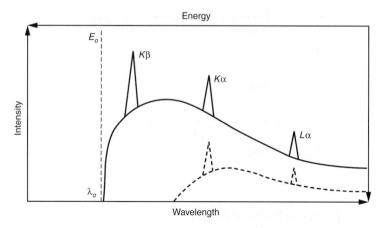

Figure A2.6 *Spectrum of X-ray emissions showing the peak emissions that are characteristic to the target element. The dotted curve illustrates the effect of reducing the energy of the incident beam.*

For example, copper is a common cathode used for the production of X-rays. The difference in energy between the shells and the corresponding frequencies of emitted X-rays is:

The implication of this second source of X-ray emission is that characteristic peaks (e.g. $K\alpha$, $K\beta$, $L\alpha$) will be superimposed on the continuous emission spectrum (Figure A2.6). The dotted line in Figure A2.6 represents the spectrum that would be obtained for an incident electron beam with lower energy. The energy needed to displace a core electron is high and therefore it is possible that the incident beam will not be sufficiently powerful to do this. The lower energy line in Figure A2.6 is unable to develop the $K\beta$ peak, but will produce the $K\alpha$ and $L\alpha$ peaks. Therefore, if a particular characteristic wavelength is desired the energy of the incident beam must be set accordingly. In reality, the situation is more complex than described above, as there are other possible shell transitions:

$$L3 \rightarrow K \quad K\alpha_1$$
$$L2 \rightarrow K \quad K\alpha_2$$
$$\text{etc.}$$

However, the energy difference between these transitions tends to be small.

Applications
The wavelengths and energy levels of the emissions from this type of inelastic interaction are used for the production of a monochromatic X-ray for use in XRD analysis. In a complementary fashion the study of the char-

acteristics of the X-ray fluorescence and the Auger electron emissions can be used to identify elements present in a sample of material. These methods are discussed at greater length in Chapter 6.

X-ray–solid interactions

In general, the interactions between X-rays and a solid are much weaker than those between electrons and a solid. Again, they can be subdivided into elastic and inelastic interactions.

Elastic interactions

The X-ray photons interact elastically with the electrons of the atoms and undergo a scattering in the solid. A beam of X-rays impinging on the surface of a solid with a regular atomic lattice structure undergoes diffraction, an effect quantified by Bragg's law. XRD enables the identification of crystal phases. This effect is the basis of the analytical method of XRD, which is described in greater detail in Chapter 6.

Inelastic interactions

If a photon of X-ray gives all its energy to an orbital electron in one interaction, it is possible to elevate the energy level of the electron or even ionize the atom by expulsion of the electron. The subsequent relaxation of the atom results in the emission of a photon of X-ray or an electron. This is parallel to the case discussed under electron–solid interactions. To enable the displacement of a core electron a high-energy photon is required; thus, the X-ray source used in this type of analysis produces hard X-rays (short wavelength). The X-ray fluorescence is characteristic of the element and therefore the measurement of the energy levels, or wavelength, of these emissions can be used to analyse chemical composition.

Appendix 3

Structure and description of crystals

Many of the compounds encountered in the study of concrete and its deterioration are crystalline. The aim of this appendix is to provide the necessary background to understand the terminology used in the interpretation of the analysis of solid structures. It contains a résumé of the following areas:

- atomic structure of crystals and amorphous substances
- Miller indices, $(h k l)$ or $[s t u]$
- description of crystals.

For background reading, see references 1–3.

Crystal structure

Crystalline and amorphous forms

A solid state can present in either a crystal form or an amorphous form. The crystalline solid state can be characterized by an ordered structure: the position of each atom is defined within a three-dimensional order. Frequently, the growth of a crystal permits the development of a polyhedral figure limited by plane faces; for example, the formation of quartz crystals in a geode. In the case of an amorphous solid only the distance between atoms is fixed by an ionic or covalent bonding. Glass is a good example of an amorphous compound. The crystal structure is not necessarily linked to the external form of the crystal. Ground quartz has an irregular form; however, the atomic arrangement, within the particles, is always the same. The smallest atomic arrangement of atoms from which a crystal structure can be constructed is known as a **unit cell**. If one pictures a lump of sugar as being a unit cell, a three-dimensional arrangement of these lumps need not have the same shape as the elemental cell. Two crystal forms of iron sulfide (FeS_2) are shown in Figure A3.1; the unit cell of pyrite is cubic and marcasite is orthorhombic.

The anisotropy found in the position of atoms in a unit cell translates into the physical properties of a crystalline material, which will consequently

Figure A3.1 *SEM image of two crystal forms of iron sulfide, FeS_2, marcasite (left, $\times 2500$) and pyrite (right, $\times 150$). Pyrite occurs as a cubic or pyritohedral (12-sided polyhedral with pentagonal faces) structure, whereas marcasite is orthorhombic.*

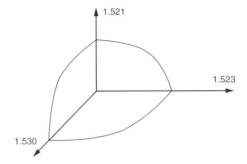

Figure A3.2 *Variation of the refractive index of gypsum with orientation.*

be anisotropic. Many crystals have refractive indices that vary with orientation. This **optical anisotropy** is used for crystal identification in petrography. For a given crystal (e.g. gypsum) one can note the refractive indices with respect to the atomic orientation. Figure A3.2 shows a representation of the refractive index as an eighth of an ellipsoid.

Another example of this anisotropy is seen in the coefficient of thermal expansion. In calcite, for example, this coefficient is positive in one direction and negative in another. This effect explains the frost susceptibility of marble, which is a metamorphic limestone in which crystals are assembled in different orientations. Figure A3.3 shows two adjoining crystals: if one crystal expands with a change in temperature and the other contracts, there will be a possible rupture between the two crystals. Another example of this effect is seen in sawn granite slabs. This material cannot be subjected to high temperatures, as in this case there is an anisotropy between crystal species at different orientations, and there will also be an anisotropy between crystals of different species.

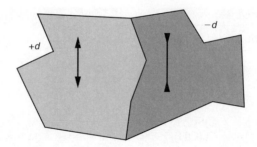

Figure A3.3 *Anisotropic crystal expansion.*

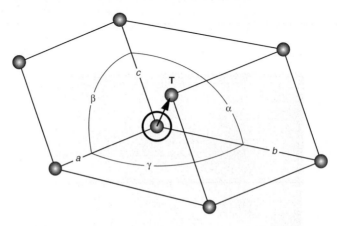

Figure A3.4 *The unit cell.*

A crystal structure can be described by the shape and size of a unit cell of the crystal lattice. This cell maintains the structure of the entire lattice. A unit cell can be described by a group of points (Figure A3.4), the lengths of the three sides being *a*, *b* and *c* and the interaxial angles between the faces α, β and γ. In this figure the distance and orientation between points can be represented vectorially, e.g.

$$\mathbf{OT} = \mathbf{a} + \mathbf{b} + \mathbf{c}$$

Table A3.1 shows various forms of unit cell. The dimensions *a*, *b* and *c* are known as the **interatomic distance**. If the position of atoms in a solid is fixed in a tridimensional lattice, planes that contain a large number of atoms can be identified. These are defined as **atomic planes** and the distance between these parallel planes of atoms is known as the **interplanar distance**. On the basis of a cubic unit cell with interatomic distance *a*, the interplanar distances would be *a*, 2*a*, 3*a*, ..., etc. These dimensions can be measured by techniques such as X-ray diffraction (XRD) and are expressed in terms of Ångström units as nanometres (1 nm = 10 Å).

Table A3.1   *Unit cells*

Crystal description	Examples	Size	Angles
Cubic	Calcium oxide C$_3$A, pyrite	$a = b = c$	$\alpha = \beta = \gamma = 90°$
Tetragonal or quadratic	ZnO	$a = b \neq c$	$\alpha = \beta = \gamma = 90°$
Orthorhombic	C$_2$S(γ) Marcasite	$a \neq b \neq c$	$\alpha = \beta = \gamma = 90°$
Hexagonal	Ca(OH)$_2$	$a = b \neq c$	$\alpha = \beta = 90°, \gamma = 120°$
Rhombohedral or trigonal	Quartz calcite	$a = b = c$	$\alpha = \beta = \gamma \neq 90°$
Monoclinic	Gypsum feldspar	$a \neq b \neq c$	$\alpha = \gamma = 90°, \beta \neq 90°$
Triclinic	C$_3$S	$a \neq b \neq c$	$\alpha \neq \beta \neq \gamma \neq 90°$

Figure A3.5   *SEM image of gypsum crystals (×190). The needle-like gypsum crystal has been subjected to a shock. The resulting cleavage planes give the crystal its laminated appearance.*

Cleavage

If the atomic bonding forces between atomic planes are weak, e.g. a large distance between two successive planes or where the planes carry the same sign electric charge, the crystal will have a tendency to split in this direction; this is known as the cleavage plane. Lime, for example, has a cleavage plane, the distance between successive atomic planes being 4.90 Å with weak binding forces. Observation by scanning electron microscopy (SEM) often shows the platelet structure of such materials. An example of a cleavage plane is shown in Figure A3.5.

Miller indices

A crystal structure, a lattice, can be characterized by a spatial network of node points set out in a regular fashion. At each of these nodes there is an atom or a group of atoms. The lattice is obtained by the translation of a vector in the unit cell (see Figure A3.4):

$$\mathbf{OT} = m\mathbf{a} + p\mathbf{b} + q\mathbf{c}$$

where m, p and q are integers.

This equation relates to a vector from a node at the origin, point O, to a node T where the same atom, or group of atoms, can be found. It is often necessary to describe directions and planes within a crystal lattice, which can be defined by a system of Miller indices.

Atomic directions

A direction in a lattice is expressed in terms of a vector joining two identical atoms. The coordinate of a point with respect to an origin O will be x,y,z. This coordinate system is transformed to express the values in terms of the lengths a, b and c:

$$\frac{x}{a}, \quad \frac{y}{b}, \quad \frac{z}{c}$$

s, t and u are Miller indices for the atomic direction. These numbers are co-prime integers, i.e. the lowest number in the set is a prime number.

The Miller index of direction is always expressed as a vector enclosed in square brackets [s t u].

Example: find the Miller index of the atomic direction AB shown in Figure A3.6.

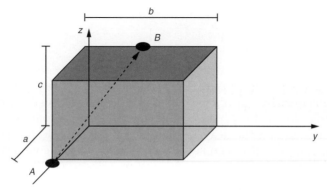

Figure A3.6 *Miller index of a direction.*

Coordinates of A $(x, y, z) = (a, 0, 0)$

Transposed coordinates $A = \left(\dfrac{x}{a}, \dfrac{y}{b}, \dfrac{z}{c} \right) = \left(\dfrac{a}{a}, \dfrac{0}{b}, \dfrac{0}{c} \right)$ or $1, 0, 0$

Coordinates of B $(x, y, z) = \left(0, \dfrac{b}{2}, c \right)$

Transposed coordinates $B = \left(\dfrac{0}{a}, \dfrac{1}{2b}, \dfrac{1}{c} \right)$ or $0, \dfrac{1}{2}, 1$

Transposed coordinates of A (tail) $= 1, 0, 0$
Transposed coordinates of B (head) $= 0, \frac{1}{2}, 1$
Subtracting head from tail:

$$0, \frac{1}{2}, 1 - 1, 0, 0 = -1, \frac{1}{2}, 1$$

Fractions are eliminated by multiplying by 2:

$$2(-1, \frac{1}{2}, 1) = -2, 1, 2$$

Negative signs are written with the convention of a bar over the number:
Miller indices of the direction AB $[\bar{2}\,1\,2]$

A more elegant approach is to express the problem vectorially.
The coordinates are transformed to A' and B'

$$OA' = \begin{pmatrix} 1 \\ 0 \\ 0 \end{pmatrix} \qquad OB' = \begin{pmatrix} 0 \\ \frac{1}{2} \\ 1 \end{pmatrix}$$

Consider a vector triangle $OA'B'$

$$OA' + A'B' = OB'$$

$$A'B' = \begin{pmatrix} 0-1 \\ \frac{1}{2}-0 \\ 1-0 \end{pmatrix} = \begin{pmatrix} -1 \\ \frac{1}{2} \\ 1 \end{pmatrix}$$

The vector $A'B'$ is multiplied by 2 to obtain a co-prime integer set. Using the same notation as set out above, the Miller index of the direction AB is $[\bar{2}\,1\,2]$.

Atomic planes

A plane can be defined within a crystal lattice by the use of a Miller index. The indices defining a plane, three co-prime integers h, k and l, are written within round brackets ( ) to differentiate them from the direction indices; a family of planes is described by the notation $\{h\,k\,l\}$.

A set of planes in a lattice is indexed with three Miller indices, h, k and l. These indices specify the direction of a plane with respect to a unit cell. Assuming the side lengths of a monoclinic unit cell are a along the x axis, b along the y axis and c along the z axis, this coordinate system is transformed to express values in terms of the lengths a, b and c:

$$x = a/h', \quad y = b/k', \quad \text{and } z = c/l'$$

A given plane intersects the axis at the points $n\mathbf{a}$, $m\mathbf{b}$ and $p\mathbf{c}$. Thus,

$$n\mathbf{a} = \frac{\mathbf{a}}{h'} \quad m\mathbf{b} = \frac{\mathbf{b}}{k'} \quad p\mathbf{c} = \frac{\mathbf{c}}{l'}$$

Therefore

$$(h'\,k'\,l') = \left(\frac{1}{n} \quad \frac{1}{m} \quad \frac{1}{p} \right)$$

The indices are always expressed as co-prime integers (hkl)

$$(hkl) \propto \left(\frac{mp}{mnp} \quad \frac{np}{mnp} \quad \frac{mn}{mnp} \right)$$

$$(hkl) = (mp \quad np \quad mn)$$

Once in this form the Miller index is expressed as the lowest common multiple.

Example: find the Miller index of the plane shown in Figure A3.7.

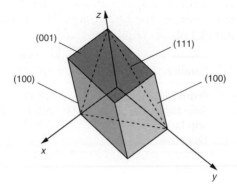

Figure A3.7 *Definition of a plane using Miller indices.*

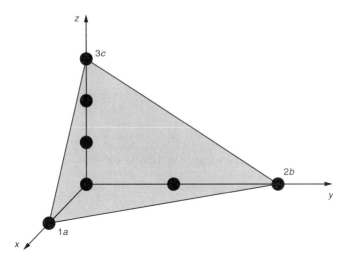

Figure A3.8

The intersections of the x, y and z axis are a, $2b$ and $3c$, respectively (Figure A3.8); $n = 1$, $m = 2$, $p = 3$. Therefore, taking the reciprocal,

$$\begin{pmatrix} \dfrac{1}{1} & \dfrac{1}{2} & \dfrac{1}{3} \end{pmatrix}$$

$$\begin{pmatrix} \dfrac{6}{6} & \dfrac{3}{6} & \dfrac{2}{6} \end{pmatrix}$$

$$(hkl) = (6 \quad 3 \quad 2)$$

It can be seen from the above example that if a plane with the intercepts $2a$, $4b$ and $6c$ were chosen, the calculation would lead to the values of (**12 6 4**). The protocol for expressing Miller indices dictates that the values must be expressed as the lowest multiples, i.e. (**6 3 2**). The plan chosen for the analysis must belong to the family of equidistant planes but does not have to be the closest to the origin. The Miller index for an atomic plane represents a family of parallel planes within the lattice.

Parallelepiped	Miller indices
Front face	(1 0 0)
Side face	(0 1 0)
Top face	(0 0 1)
Diagonal	(1 1 1)

The rectangular coordinate system does not readily lend itself to hexagonal structures and therefore a variant on the above system of indices is used; these are known as the Miller–Bravais indices.

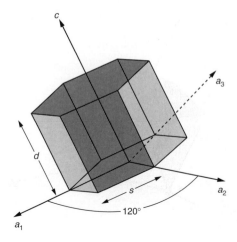

Figure A3.9

Consider the hexagonal coordinate system illustrated in Figure A3.9. As in the previous case the indices are defined in terms of the intercepts of the axis; for a hexagonal side of s and height d the indices can be defined by the intercepts s/h s/k s/i d/l. Considering the three axes at a_1, a_2 and a_3, from inspection of a hexagon it is evident that if a_1 and a_2 are known, a_3 is automatically defined.

$$i = -(h + k)$$

For this reason, when the indices for a hexagonal structure $(hkil)$ are quoted the **i** index is often replaced by a dot, e.g. (21.8)

Consider the diagonal plane shown in Figure A3.10. The intercepts of the four axes are:

Axis	Intercept	Transformed coordinates	Miller indices
a_1	s	$h = s/s$	1
a_2	s	$k = s/s$	1
a_3	$-s/2$	$i = s/(-s/2)$	$\bar{2}$
c	$d/2$	$c = d/(d/2)$	2

The coordinates of the intercepts are transformed as in the previous case, i.e.

$$h = \frac{s}{\text{intercept of } a_1 \text{ axis}}, \text{ etc.}$$

Therefore, the Miller indices for the planes are $(1\ 1\ \bar{2}\ 2)$ or $(1\ 1\ .\ 2)$. The indices for the other faces are shown in Figure A3.11.

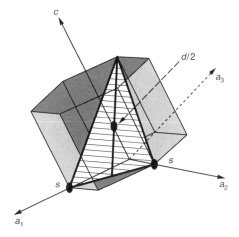

Figure A3.10   *The hatched plane in the hexagonal cell intersects the a_1 axis at s, the a_2 axis at s, the a_3 axis at $-s/2$ and the c axis at $d/2$.*

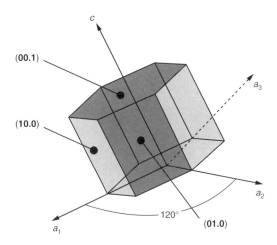

Figure A3.11

Description of a crystal lattice

A unit cell of a crystal lattice can be described in terms of three vectors, **a, b** and **c**. A plane within the lattice can be described by the lattice intercepts m, n and p along the three axis **a, b** and **c** (see Figure A3.12). Reciprocals of these lattice intercepts are then taken:

$$h' = 1/m \quad k' = 1/n \quad \text{and } l' = 1/p.$$

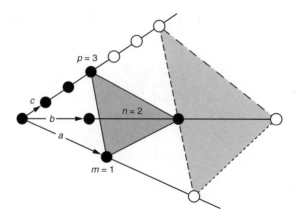

Figure A3.12

These values are multiplied by an integer to obtain three co-prime integers *h*, *k* and *l*. These are the Miller indices of the plane (*hkl*).

In Figure A3.12 a plane is defined by the lattice intercepts $m = 1$, $n = 2$, $p = 3$.

$$h' = 1/1 \quad k' = 1/2 \quad \text{and } l' = 1/3$$

Multiplying by an integer of 6, three co-prime numbers are obtained, thus giving the Miller indices of the plane (**6 3 2**). It can be seen that parallel planes exist such as:

$$m = 2 \quad n = 4 \quad p = 6.$$

It follows that the co-prime numbers of all these planes will be the same (**6 3 2**).

The interplanar distance d_{hkl} is the perpendicular distance between parallel planes described by the Miller index (*hkl*). In 1913 P.P. Eward introduced the concept of a reciprocal space for calculating interplanar distance. This notion is based on the construction of a reciprocal lattice. This space is defined by vectors **A**, **B** and **C** which are the reciprocals of the vectors **a**, **b** and **c** that define the real unit cell and lattice. The reciprocal lattice is defined by the vector $\mathbf{G}_{hkl}$.

$$\mathbf{A} = \sigma^2 \frac{\mathbf{b} \times \mathbf{c}}{\mathbf{a} \cdot (\mathbf{b} \times \mathbf{c})} \quad \mathbf{B} = \sigma^2 \frac{\mathbf{c} \times \mathbf{a}}{\mathbf{b} \cdot (\mathbf{c} \times \mathbf{a})} \quad \mathbf{C} = \sigma^2 \frac{\mathbf{a} \times \mathbf{b}}{\mathbf{c} \cdot (\mathbf{a} \times \mathbf{b})}$$

where σ is a constant, and **a**, **b** and **c** are the vectors defining the unit cell.

The distance between two nodes in this reciprocal lattice will be given by the equation:

$$\mathbf{G} = k\mathbf{A} + h\mathbf{B} + l\mathbf{C}$$

where *h*, *k* and *l* are the Miller indices.

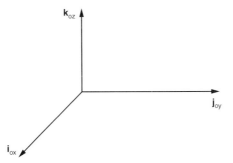

Figure A3.13

The interplanar distance in the lattice network can then be found:

$$d_{hkl} = \frac{\sigma^2}{\|G\|}$$

Example: find the interplanar distance for a plane *hkl* for a orthorhombic lattice.

Within the orthorhombic lattice the three sides of the unit cell are equal to a, b and c and the directions are defined by the unit vectors $\mathbf{i}_{ox}$ $\mathbf{j}_{oy}$ and $\mathbf{k}_{oz}$ at right angles (Figure A3.13).

The reciprocal lattice **A**, **B** and **C** can now be defined:

$$\mathbf{A} = \sigma^2 \frac{\mathbf{i}_{ox}}{a} \quad \mathbf{B} = \sigma^2 \frac{\mathbf{j}_{oy}}{b} \quad \mathbf{C} = \sigma^2 \frac{\mathbf{k}_{oz}}{c}$$

It can be seen from the above that the reciprocal vector **A** is parallel to the lattice vector **a**.

$$\mathbf{j}_{oy} \times \mathbf{k}_{oz} = \mathbf{i}_{ox}$$

The reciprocal lattice vector is therefore given by:

$$G = \sigma^2 \left(\frac{h\mathbf{i}_{ox}}{a} + \frac{k\mathbf{j}_{oy}}{b} + \frac{l\mathbf{k}_{oz}}{c} \right)$$

Thus, an expression can be written for the interplanar distance d_{hkl} of a cubic lattice:

$$\|G\| = \sigma^2 \sqrt{\left(\frac{h^2}{a^2} + \frac{k^2}{b^2} + \frac{l^2}{c^2} \right)}$$

Therefore, the distance to the nearest parallel plane is:

$$d_{hkl} = \frac{\sigma^2}{\|G\|}$$

$$d_{hkl} = \frac{1}{\sqrt{\left(\dfrac{h^2}{a^2} + \dfrac{k^2}{b^2} + \dfrac{l^2}{c^2}\right)}}$$

This approach enables the determination of interplanar spacing between any parallel planes in a real lattice.

References

1 Battey M.H. and Pring A. *Mineralogy for students*, 3rd edn. Longman, London, 1997.
2 Askeland D.R. *The science and engineering of materials*, 3rd edn. Chapman and Hall, London, 1996, Chs 2 and 3.
3 Regourd M. L'hydratation du ciment Portland. In *Le béton hydraulique*. Presses de l'École Nationales des Ponts et Chaussées, Paris, 1982, pp. 196–198.

Appendix 4

Mineralogical data

Interplanar distances for X-ray diffraction (XRD) are given in Ångström units. The three strongest diffraction peaks are noted in the tables. Where appropriate, the JCPDS card reference has been quoted.

C   CaO
S SiO_2
A Al_2O_3
s SO_4
M   MgO

Crystal form: the crystal form given in the tables is found if the crystallization takes place under ideal condition. However, in practice, the form of the crystals may be variable.

Cements

Mineral	XRD data						Crystal form	Unit cell
Alite (C$_3$S)	2.776	vs	2.602	vs	2.185	vs	Equant grains	Triclinic
Beta Belite (βC$_2$S)	2.794	vs	2.785	vs	2.748	s	Ill defined, often twinned	Monoclinic
Brownmillerite (C$_4$AF)	2.63	vs	2.77	s	1.92	s	Prisms	Orthorhombic
Tricalcium aluminate (C$_3$A)	2.70	vvs	1.908	s	4.08	ms	Equant grains	Cubic
Gypsum (CaSO$_4$·2H$_2$O)	7.56	vs	4.27	s	3.06	s	Tablets (010) cleavage	Monoclinic
Free lime (CaO)	2.4	vvs	1.701	s	2.778	ms	Cubic	Cubic
Monocalcium aluminate (CA)	14.3	vs	7.16	vs	3.56	vs	Irregular grains	Monoclinic
Periclase (MgO)	2.106	vvs	1.489	ms	2.431	w	Cubes	Cubic
Merwinite (C$_3$MS$_3$)	2.687	vvs	2.671	ms	2.653	s	Granular or prismatic	Monoclinic
Melilite gehlenite (C$_2$AS)	2.872	vs	2.478	w	1.761	w	Prisms or tablets	Tetragonal
åkermanite (C$_2$MS$_2$)	2.845	vs	2.431	w	1.754	w		

Equant: equidimensional.

Hydration products

Mineral	XRD data						Crystal form	Unit cell
Calcium silicate hydrates								
CSH II	3.07	vvs	9.8	vs	2.8	vs	Fibre bundles (semi-crystalline)	Orthorhombic
CSH I	12.56	vs	3.07	vs	2.8	s	Crumpled foils (semi-crystalline)	Orthorhombic
11Å tobermorite	3.07	vvs	11.0	vs	2.97	s	Elongated plates or lathes	Orthorhombic
9Å tobermorite ($C_5S_6H_{2.5}$)	3.04	v	9.67	s	4.83	s		Orthorhombic
Other hydrates								
Portlandite (CH)	2.628	vvs	4.9	vs	1.927	ms	Hexagonal flakes (00.1) cleavage	Triagonal
Ettringite (C_3AsH_{32})	9.73	vvs	5.61	vs	2.56	s	Hexagonal plates or needles	Hexagonal
Monosulfate (C_3AsH_{12})	8.9	vvs	4.45	vs	2.190	s	Hexagonal plates (00.1) cleavage	Hexagonal
Aluminous cement hydrates								
Monocalcium aluminate (CAH_{10})	14.3	vs	7.16	vs	3.56	s	Crumpled foils	–
Gibbsite $Al(OH)_3$	4.85	vvs	4.37	vs	4.32	s	Tablets or prisms (001) cleavage	Monoclinic
Hydrogarnet (C_3AH_6)	2.3	vs	2.04	vs	5.13	s	Trapezohedral or cubic	Cubic
Dicalcium aluminate (C_2AH_8)	10.7	vvs	2.87	vvs	2.55	s	Hexagonal plates (00.1) cleavage	Hexagonal

Reaction products

Mineral	XRD data						Crystal form	Unit cell
Calcite (CaCO₃)	3.035	vvs	2.285	s	2.095	s	Rhombs (10.4) cleavage	Hexagonal or rhombic
Vaterite (μ-CaCO₃)	3.58	s	3.30	s	2.73	s	Fibrous	Hexagonal or trigonal
Thaumasite ([3Si(OH)₆· 12H₂O] (SO₄)(CO₃))	9.56	vvs	5.51	s	3.41	ms	Needles overgrowth on ettringite	Hexagonal
Ettringite (C₃AsH₃₂)	9.73	vvs	5.61	vs	2.56	s	Hexagonal plates or needles	Hexagonal
Gypsum (CaSO₄·2H₂O)	7.56	vs	4.27	s	3.06	s	Tablets (010) cleavage	Monoclinic
Brucite (Mg(OH)₂)	2.365	vvs	4.77	vvs	1.794	vs	Hexagonal flakes	Trigonal
Chloroaluminate hydrate (C₃A CaCl₂ xH₂O)							Hexagonal plates	

Aggregate

Mineral	XRD data						Crystal form	Unit cell
Quartz (SiO₂)	3.35	vvs	4.24	s	1.81	s	Equant grains	Trigonal
Flint								Amorphous
Calcite (CaCO₃)	3.035	vvs	2.285	s	2.095	s	Rhombs	Rhombic
Dolomite (CaMg(CO₃)₂)	2.883	vs	2.191	s	4.11	m		Rhombic
Pyrite (FeS₂)	1.629	vs	2.696	s	2.417	s	Cube	Cubic
Pozzolans								
Mullite (A₃S₂)	2.206	vs	3.428	vs	2.260	vvs	Fine needles on PFA nodules	Orthorhombic

Index